Dictionary of GENETICS

M. Follower

RAJAT PUBLICATIONS
New Delhi

Dictionary of Genetics
by *M. Follower*

This edition is published by Rajat Publications
4740/23, Ansari Road Daryaganj,
New Delhi - 110002

ISBN: 978-81-7880-299-2

Edition 2025

Printed and bound in India

RAJAT
PUBLICATIONS

4740/23, Ansari Road, Daryaganj,
New Delhi- 110002 (India)
Mob:- +91-9811484973, +91-9811105389
E-mail: rajat_publication@yahoo.com

Preface

Genetics is the area of biological study concerned with heredity and with the variations between organisms that result from it. It demands an understanding of numerous terms, such as DNA (deoxyribonucleic acid), a molecule in all cells that contains blueprints for genetic inheritance; genes, units of information about particular heritable traits, which are made from DNA; and chromosomes, DNA-containing bodies, located in the cells of most living things, that hold most of the organism's genes. The vocabulary of genetics goes far beyond these three terms, as we shall see, but these are the core concepts. Among the areas in which genetics is applied is forensic science, or the application of science to matters of law—specifically, through "DNA fingerprinting," whereby samples of skin, blood, semen, and other materials can be used to prove or disprove a suspect's innocence. Another fascinating application of genetics is the Human Genome Project, an effort whose goals include the location and identification of every gene in the human body.

This dictionary includes a vast range of terms and concepts dealing with biochemistry, cell and developmental biology, immunology, hereditary diseases, and molecular evolution, as well as advances in genomics and proteomics.

—Editor

# #

2μ (2 micron) plasmid. A naturally occurring extragenomic (not part of the genome) circular DNA molecule found in some yeast cells, with a circumference of 2μ. Engineered to form the basis for several types of gene vectors in yeast.

3' end/5' end. A nucleic acid strand is inherently directional, and the "5 prime end" has a free hydroxyl (or phosphate) on a 5' carbon and the "3 prime end" has a free hydroxyl (or phosphate) on a 3' carbon (carbon atoms in the sugar ring are numbered from 1' to 5'). That's simple enough for an RNA strand or for single-stranded (ss) DNA. However, for double-stranded (ds) DNA it's not so obvious - each strand has a 5' end and a 3' end, and the 5' end of one strand is paired with the 3' end of the other strand (it is "antiparallel"). One would talk about the 5' end of ds DNA only if there was some reason to emphasize one strand over the other - for example if one strand is the sense strand of a gene. In that case, the orientation of the sense strand establishes the direction.

3' flanking region. A region of DNA which is NOT copied into the mature mRNA, but which is present adjacent to 3' end of the gene. It was originally thought that the 3' flanking DNA was not transcribed at all, but it was discovered to be transcribed into RNA, but quickly removed during processing of the primary transcript to form the mature mRNA. The 3' flanking region often contains sequences which affect the formation of the 3' end of the message. It may also contain enhancers or other sites to which proteins may bind.

3' untranslated region. A region of the DNA which IS transcribed into mRNA and becomes the 3' end or the message, but which does not contain protein coding sequence. Everything between the stop codon and the polyA tail is considered to be 3' untranslated. The 3' untranslated region may affect the translation efficiency of the mRNA or the stability of the mRNA. It also has sequences which are required for the addition of the poly(A) tail to the message (including one known as the "hexanucleotide", AAUAAA).

5' flanking region. A region of DNA which is NOT transcribed into RNA, but rather is adjacent to 5' end of the gene. The 5'-flanking region contains the promoter, and may also contain enhancers or other protein binding sites.

5' untranslated region. A region of a gene which IS transcribed into mRNA, becoming the 5' end of the message, but which does not contain protein coding sequence. The 5'-untranslated region is the portion of the DNA starting from the cap site and extending to the base just before the ATG translation initiation codon. While not itself translated, this region may have sequences which alter the translation efficiency of the mRNA, or which affect the stability of the mRNA.

5' untranslated region. The short sequence between the transcription initiation site and the start of translation that is retained in mRNA but not translated. It contains the ribosomal binding site (leader sequence) and signal sequence.

A

A (aminoacyl) site. The site on the ribosome occupied by an aminoacyl-tRNA just prior to peptide bond formation.

A form DNA. The form of DNA at high humidity; it has tilted base pairs and more base pairs per turn than does B DNA.

A. Adenine or adenosine.

ablation experiment. An experiment designed to produce an animal deficient in one or a few cell types, in order to study cell lineage or cell function. The idea is to make a transgenic mouse with a toxin gene (often diphtheria toxin) under control of a specialized promoter which activates only in the target cell type. When embryo development progresses to the point where it starts to form the target tissue, the toxin gene is activated, and that specific tissue dies. Other tissues are unaffected.

abortive transduction. The failure of a transducing DNA segment to be incorporated into the recipient chromosome.

acentric chromosome. A chromosome having no centromere.

acentric fragment. A chromosomal piece without a centromere.

achondroplasia. A type of dwarfism in humans inherited as an autosomal dominant phenotype.

acrocentric chromosome. (1) A chromosome whose centromere lies very near one end. See also metacentric chromosome. (2) A chromosome with its centromere towards one end. Human chromosomes 13,14,15,21,22 are acrocentric.

acrylamide gels. A polymer gel used for electrophoresis of DNA or protein to measure their sizes (in daltons for proteins, or in base pairs for DNA). See "Gel Electrophoresis". Acrylamide gels are especially useful for high resolution separations of DNA in the range of tens to hundreds of nucleotides in length.

activation energy (deltaG). Energy needed to initiate a chemical reaction.

active site. The part of a protein that must be maintained in a specific shape if the protein is to be functional, for example, the part to which the substrate binds in an enzyme. The part of an enzyme where the actual enzymatic function is performed.

adaptation. (1) Adjustment to environmental demands through the long-term process of natural selection acting on genotypes. (2) In the evolutionary sense, some heritable feature of an individual's phenotype that improves its chances of survival and reproduction in the existing environment.

adaptive immunity. A collective term for the long-lasting and specific response of lymphocytes to antigens. Requires the MHC, T-cell receptors (TCR) and immunoglobulins (Ig) as well as enzymes with a recombinase activity (for the rearrangements at TCR and Ig gene loci). Present in all vertebrates except jawless fish (see innate immunity).

adaptive landscape. The surface plotted in a three-dimensional graph; with all possible combinations of allele frequencies for different loci plotted in the plane; and mean fitness for each combination plotted in the third dimension.

adaptive peak. A high point (perhaps one of several) on an adaptive landscape; selection tends to drive the genotype composition of the population toward a combination corresponding to an adaptive peak.

adaptive surface. See adaptive landscape.

adaptive value. See fitness.

additive and non-additive components. In studies of heredity, the portions of the genetic component that are passed and not passed to offspring, respectively.

additive genetic variance. Genetic variance associated with the average effects of substituting one allele for another.

additive model. A mechanism of quantitative inheritance in which alleles at different loci either add a fixed amount to the phenotype or add nothing.

adenine. A purine base that pairs with thymine in the DNA double helix. See purines.

adenosine triphosphate. See ATP.

adenosine. The nucleoside containing adenine as its base.

adjacent segregation. In a reciprocal translocation heterozygote during meiosis the segregation of a translocated and a normal chromosome together, giving unbalanced gametes with duplications and deficiencies leading to non-viable zygotes. Adjacent segregation is of two kinds depending on whether non-homologous (adjacent-1) or homologous (adjacent-2) centromeres segregate together. Adjacent-1 segregation is the usual type of adjacent segregation, adjacent-2 segregation is rare.

adjacent-1 segregation. Segregation of non-homologous centromeres during meiosis in a reciprocal translocation heterozygote such that unbalanced gametes with duplications and deficiencies are produced. See alternate segregation and adjacent-2 segregation.

adjacent-2 segregation. Segregation of homologous centromeres during meiosis in a translocation heterozygote such that unbalanced gametes with duplications and deficiencies are produced. See alternate segregation and adjacent-1 segregation.

ADP. Adenosine diphosphate.

advanced (synonym: derived; opposite: primitive). In phylogenic studies, an organism or character further removed from an evolutionary divergence than a more primitive one.

affected. Individuals in a pedigree that exhibit the specific phenotype under study.

agarose gels. A polysaccharide gel used to measure the size of nucleic acids (in bases or base pairs). See "Gel Electrophoresis". This is the gel of choice for DNA or RNA in the range of thousands of bases in length, or even up to 1 megabase if you are using pulsed field gel electrophoresis.

agnatha (means jawless). The Class Agnatha represents the most primitive 'jawless' vertebrates. MHC genes have been cloned from all vertebrate classes except Agnatha.

agrobacterium tumefaciens. A soil bacterium that causes a cancer-like plant disease (crown gall) in dicotyledenous plants (all agricultural crops except cereals). It contains the Ti plasmid. The tumor induction ability of the bacterium spreads to neighboring cells via the plasmid.

Ala. Alanine (an amino acid).

albino. A pigmentless white phenotype, determined by a mutation in a gene coding for a pigment-synthesizing enzyme.

algae. A heterogeneous group of aquatic, unicellular, colonial or multicellular, eukaryotic and photosynthetic organisms. They belong to the Kingdom Protista and include the multicellular red (rhodophyte), green and brown (kelp) algae. They are not plants but all land plants evolved from the green algae (see also Chromista).

alkylating agent. A chemical agent that can add alkyl groups (for example, ethyl or methyl groups) to another molecule; many mutagens act through alkylation.

allee effect. The benefit individuals gain from the presence of conspecifics. Link to a brief explanation of **Allee effect**.

allele frequency. Often called gene frequency. A measure of how common an allele is in a population; the proportion of all alleles at one gene locus that are of one specific type in a population.

allele. A known variation (version) of a particular gene. Formerly called allelomorph.

allele. Alternative form of a gene. One of the different forms of a gene that can exist at a single locus.

allelic association. See linkage disequilibrium.

allelic exclusion. (1) A process whereby only one immunoglobulin light chain and one heavy chain gene are transcribed in any one cell; the other genes are repressed. (2) Expression of only one of the two homologous alleles at a locus in the case of heterozygosity. This usually occurs at loci such as immunoglobulin or T cell receptor (TCR) genes where a functional rearrangement among genes takes place. One of the alleles is either non-functionally or incompletely rearranged and not expressed. This way, each T-cell expresses only one set of TCR genes.

allelopathy. The influence exerted by a living plant on other plants nearby or microorganisms through production of a chemical.

allogeneic. Two genetically dissimilar individuals of the same species like any two human beings except monozygotic twins.

allopatric speciation. Speciation in which the evolution of reproductive isolating mechanisms occurs during physical separation of the populations.

allophenic. Chimeric, i.e., composed of cells of two different genotypes (also called hybrid).

allopolyploid. Polyploid produced by the hybridization of two species. See amphidiploid.

allorecognition. Recognition by T cells of the **MHC** molecules on an allogeneic individual's antigen-presenting cells which results in allograft rejection *in vivo* and mixed lymphocyte reaction (MLR) *in vitro*.

allosteric protein. A protein whose shape is changed when it binds a particular molecule. In the new shape the protein's ability to react to a second molecule is altered.

allosteric transition. A change from one conformation of a protein to another.

allotype. The protein product (or the result of its activity) of an allele which may be detected as an antigen in another member of the same species (e.g., histocompatibility antigens, immunoglobulins), obeying the rules of simple Mendelian inheritance.

allozygosity. Homozygosity in which the two alleles are alike but unrelated. See autozygosity.

allozyme. Form of an enzyme that differs in amino acid sequence, as shown by electrophoretic mobility or some other property, from other forms of the same enzyme and is encoded by one allele at a single locus. See isozymes.

α-helix. Common secondary structure of proteins in which the linear sequence of amino acids is folded into a spiral that is stabilized by hydrogen bonds between the carboxyl oxygen of each peptide bond.

altered self. A term used to describe the MHC molecule associated with a peptide rather than in its native form. Thus, a native MHC molecule does not induce an immune reaction except when it is presenting a peptide.

alternate segregation. At meiosis in a reciprocal translocation heterozygote, the passage (segregation) of both normal chromosomes to one pole and both translocated chromosomes to the other pole, giving genetically balanced gametes. Segregation of centromeres during meiosis in a reciprocal translocation heterozygote such that genetically balanced gametes are produced.

alternation of generations. (1) An alternation of sexual (haploid) and asexual (diploid) form of generations in a life cycle (example: aphids). The relative dominance of each phase is variable in each organism (mosses have a dominant haploid phase whereas angiosperms have a dominant diploid phase). Besides aphids, Daphnia (water flea), rotifers, Hydra have alternation of generations in response to environmental conditions. (2) The alternation of gametophyte and sporophyte stages in the life cycle of a plant.

alternative splicing. Formation of diverse mRNAs through differential splicing of the same RNA precursor.

alternative splicing. Various ways of splicing out introns in eukaryotic pre-mRNAs resulting in one gene producing several different mRNAs and protein products.

altruism. (1)A form of behaviour in which an individual risks lowering its fitness for the benefit of another. (2) Helping others without direct benefit, and sometimes harm, to oneself.

Alu family. A dispersed intermediately repetitive DNA sequence found in the human genome in about three hundred thousand copies. The sequence is about 300 bp long. The name **Alu** comes from the restriction endonuclease AluI that cleaves it.

amber codon. The codon UAG, a nonsense codon.

amber suppressor. A mutant allele coding for a tRNA whose anticodon is altered in such a way that the suppressor tRNA inserts an amino acid at an amber codon in translation suppressing (preventing) termination.

Ames test. A widely used test to detect possible chemical carcinogens; based on mutagenicity in the bacterium Salmonella.

amino acid. The basic building block of proteins (or polypeptides). Containing a basic amino (NH2) group, an acidic carboxyl (COOH) group and a side chain (R - of a number of different kinds) attached to an alpha carbon atom. Thus:

```
  R
  |
NH2-C-COOH
  |
  H
```

aminoacyl-tRNA synthetase. Enzyme that attaches an amino acid to its cognate tRNA(s).

aminoacyl-tRNA. Transfer RNA (tRNA) molecule with its cognate amino acid attached.

amniocentesis. A technique for testing the genotype of an embryo or fetus in utero with minimal risk to the mother or the child.

amorph. A mutant showing the complete lack of some normal substance or structure.

amp resistance. See "Antibiotic resistance".

AMP. Adenosine monophosphate.

amphidiploid. An allopolyploid; a polyploid formed from the union of two separate chromosome sets and their subsequent doubling. An organism produced by hybridization of two species followed by chromosome doubling. An allotetraploid that appears to be a normal diploid.

amphipathic. A molecule that has both a hydrophobic and a hydrophilic part.

amplification of DNA. The production of many DNA copies from one or a few copies.

anagenesis. The evolutionary process whereby one species evolves into another without any splitting of the phylogenetic tree. See cladogenesis.

analogy. A similarity due to convergent evolution (common function) but not inheritance from a common ancestor (bat's wings and bird's wings). See also homology.

anaphase. The stage of meiosis and mitosis in which sister chromatids (meiosisII) or homologous chromosomes (mitosis and meiosisI) are separated by spindle fibers. An intermediate stage of nuclear division during which chromosomes or chromatids are pulled to the poles of the spindle.

aneuploid cell. A cell having a chromosome number that differs from the normal chromosome number for the species by a small number of chromosomes.

aneuploid. Individuals or cells exhibiting aneuploidy.

aneuploidy. The condition of a cell or of an organism that has additions or deletions of a small number of whole chromosomes from the expected balanced diploid number of chromosomes.

angiosperm. (1) Plant whose seeds are enclosed within an ovary. Flowering plants. (2) The most recently evolved and the largest group of plants whose reproductive organs are in their flowers (flowering plants). A superclass in the sperm plants (Spermatophyta) division belonging to the vascular plants (Tracheophyta) phylum of the plant kingdom. They are divided into two subclasses: Dicots (Magnoliopsida such as magnolia, dandelion, roses, violet) and Monocots (Liliopsida such as lilly, iris, orchid, grasses). Their ovules are enclosed in the carpel and pollen travels through the pollen tube to reach it. Angiosperms evolved in the Cretaceous era together with the Mammals. Angiosperms in Tree of Life.

animal breeding. The practical application of genetic analysis for development of lines of domestic animals suited to human purposes.

anisogamy. Sexual reproduction in which one sex produces sex cells much larger (egg) than those of the other (sperm).

anneal. Generally synonymous with "hybridize".

annealing. Spontaneous alignment of two complementary single polynucleotide (RNA, or DNA, or RNA and DNA) strands to form a double helix.

antagonistic pleiotropy. The effects of a gene which are beneficial early in life (i.e., increasing fitness) but deleterious later in life (no change in fitness after the reproductive age). Such genes will be maintained by selection, because by the time the gene exerts its damage, its bearers will already have had more offspring than other individuals.

anthropology. The study of human kind.

antibiotic resistance. Plasmids generally contain genes which confer on the host bacterium the ability to survive a given antibiotic. If the plasmid pBR322 is present in a host, that host will not be killed by (moderate levels of) ampicillin or tetracycline. By using plasmids containing antibiotic resistance genes, the researcher can kill off all the bacteria which have not taken up his plasmid, thus ensuring that the plasmid will be propagated as the surviving cells divide.

antibody. A protein produced by a B lymphocyte that protects the organism against an antigen. A protein (immunoglobulin) molecule, produced by the immune system, that recognizes a particular foreign antigen and binds to it; if the antigen is on the surface of a cell, this binding leads to cell aggregation and subsequent destruction.

anticoding strand. The DNA strand that forms the **template** for both the transcribed mRNA and the coding DNA strand.

anticodon. The three-base sequence in tRNA complementary to a codon on mRNA. A nucleotide triplet in a tRNA molecule that aligns with a particular codon in mRNA under the influence of the ribosome, so that the amino acid carried by the tRNA is added to a growing protein chain.

antigen. (1) A molecule whose shape triggers the production of antibodies (immunoglobulins) that will bind to the antigen. A foreign substance capable of triggering an immune response in an organism.

(2) Any macromolecule that triggers an immune response. Antigenicity depends on the ability of the peptide fragments to be presented by the MHC molecules.

antimorph. A mutant expressing some agent that antagonizes a normal gene product.

antimutator mutation. mutation of DNA polymerase that decreases the overall mutation rate.

anti-oncogene. A gene that prevents malignant (cancerous) growth and whose absence, by mutation, results in malignancy (e.g., retinoblastoma).

antiparallel. A term used to describe the opposite orientations of the two strands of a DNA double helix; the 5' end of one strand aligns with the 3' end of the other strand.

antisense DNA/RNA. Single stranded nucleic acid that is complementary to the coding/sense strand of a gene. It is then also complementary to the mRNA produced from the same gene.

antisense RNA. RNA product of mic (mRNA-interfering complementary RNA) genes that regulates another gene by base pairing with and thus preventing translation from its mRNA.

antisense strand. See anticoding strand.

antiterminator protein. A protein that, when bound at its normal attachment sites in DNA, lets RNA polymerase read through normal terminator sequences (e.g., the N- and Q-gene products of phage lambda).

AP endonuclease. Endonuclease that initiates excision repair at apurinic and apyrimidinic sites on DNA.

AP site. Apurinic or apyrimidinic site resulting from the loss of a purine or pyrimidine residue from the DNA.

AP-1 site. The binding site on DNA at which the transcription "factor" AP-1 binds, thereby altering the rate of transcription for the adjacent gene. AP-1 is actually a complex between c-fos protein and c-jun protein, or sometimes is just c-jun dimers. The AP-1 site consensus sequence is (C/G)TGACT(C/A)A. Also known as the TPA-response element (TRE). [TPA is a phorbol ester, tetradecanoyl phorbol acetate, which is a chemical tumor promoter]

apes. Species belonging to the Family Pongidae of the Order Primates: Gibbon (genus Hylobates), Orangutan (genus Pongo), Chimpanzee (genus Pan), and Gorilla (genus Gorilla). They have no tails.

apoptosis. The genetically programmed death of cells at specific times during embryonic morphogenesis and development, metamorphosis, and during cell turnover in adults. Defects in apoptosis are associated with maintenance of the transformed state and cancer. Anti-apoptotic proteins include Bcl-2 and HSP families (see also **caspase**).

aposematic coloring. Coloration that warns the predators about poisonous or distasteful nature of the organism. Usually black and yellow stripes in animal. Aposematic coloring of poisonous organisms is sometimes mimicked by others to gain advantage against predators (mimicry). (Link to an interesting article by Lev-Yadun on aposematic coloring in plants.)

arabidopsis thaliana. A small member of the mustard family (kitchen cress). It has a very small genome (130-140 Mbp), five chromosomes and contains almost no repetitive DNA. Its genome will be completely sequenced by the end of 2000. It is a plant model system of choice because of the additional advantages of short generation time (about five weeks), high seed production (up to 40,000 seeds per plant) and natural self-pollination (as opposed to natural cross-pollination in maize). It has five small chromosomes.

arboreal. Tree-living (like monkeys).

archaea. A prokaryote kingdom that has not diverged much from the ancestral prokaryote stock. Contemporary species of Archeabacteria live in extreme conditions. The three major groups are halobacteria, sulphobacteria and methanogens. All other prokaryotes are grouped in Eubacteria.

archezoa. One of the kingdom level taxa proposed by Cavallier-Smith which consists of the most ancient unicellular eukaryotes with a nucleus and rod shaped chromosome but no mitochondria or plastid, thus believed to be the intermediate stage between prokaryotes and eukaryotes. They are also used as evidence for the evolution of nucleus before the organelles. The intestinal parasite Giardia lamblia (a protist) is an example.

Arg. Arginine (an amino acid).

ARS. Autonomously replicating sequence. ARS is the origin of replication in yeast.

artificial selection. Selective evolutionary pressure imposed by humans to obtain breeds with certain features (such as breeding cows, dogs, chicken).

ascospore. A sexual spore from certain fungus species in which spores are found in a sac called an ascus. Haploid spores found in the asci of Ascomycete fungi.

ascus. In fungi, a sac that encloses a tetrad or an octad of ascospores. The sac in Ascomycete fungi that holds the ascospores.

asexual reproduction. Any form of reproduction not depending on a sexual process. It involves a single individual. Reproduction by cell division, fragmentation or budding.

asexual spore. See spore.

Asn. Asparagine (an amino acid).

Asp. Aspartate (an amino acid).

assignment test. A test that determines whether a locus is on a specific human chromosome by observation of the concordance of the locus and the specific chromosome in a panel of human-mouse hybrid cell lines containing only one or a few of the normal set (22 autosomes, X and Y) of human chromosomes.

association (genetic). Association refers to a concurrence greater than predicted by chance between a specific allele and another trait (for example, a disease) that may or may not have a genetic basis. Evaluation of association requires the study of unrelated individuals. Association studies may prove useful in identifying a genetic factor in a disease. Except when linkage disequilibrium exists, association is not due to genetic linkage and should not be confused with it.

assortative mating. (1) The mating of individuals with similar phenotypes. (2) Reproduction in which mate selection is not random but is based on physical, cultural, or religious grounds (see negative and positive assortative mating).

ataxia-telangiectasia. A disease (loss of muscle control, and reddening of the skin) in human beings caused by a defect in DNA repair mechanisms induced by ionising radiation (X-rays, beta and alpha particles, gamma rays).

ATG or AUG. The codon for methionine; the translation initiation codon. Usually, protein translation can only start at a methionine codon (although this codon may be found elsewhere within the protein sequence as well). In eukaryotic DNA, the sequence is ATG; in RNA it is AUG. Usually, the first AUG in the mRNA is the point at which translation starts, and an open reading frame follows - i.e. the nucleotides taken three at a time will code for the amino acids of the protein, and a stop codon will be found only when the protein coding region is complete.

atomic mass unit (amu or dalton). The basic unit of mass on an atomic scale. One amu or dalton is one-twelfth the mass of a carbon 12 atom (in other words, the mass of a hydrogen atom, 1.66×10^{-24} g). Therefore, there are 6.023×10^{23} amu in one gram (Avogadro number).

ATP (adenosine triphosphate). The energy molecule of cells, synthesized mainly in mitochondria and chloroplasts; energy from the breakdown of ATP drives many important reactions in the cell.

attached X. A pair of Drosophila X chromosomes joined at one end and inherited as a single unit.

attenuator stem. A configuration of the leader transcript that signals transcription termination in attenuator-controlled amino acid operons.

attenuator. A control region at the promoter end of repressible amino acid operons that exerts transcriptional control based on the translation of a small leader peptide gene. A region adjacent to the structural genes of the trp (tryptophan) operon; in the presence of tryptophan this region acts to reduce the rate of transcription from the structural genes.

australopithecus. The extinct genus of Plio-Pleistocene hominids found in South and East Africa. The evolutionary link between **apes** and humans.

autogamy. Nuclear reorganization in a single Paramecium cell similar to the changes that occur during conjugation.

automorphy. Unique derived characteristic; a trait present in only one member of a lineage or in only one lineage among many.

autonomous replication sequence (ARS). A segment of a DNA molecule necessary for the initiation of its replication; generally a site recognized and bound by the proteins of the replication system.

autopolyploid. A polyploid formed from the doubling of a single genome. Polyploidy in which all the chromosomes come from the same species.

autoradiography. A process by which radioactive materials, often though not exclusively incorporated into cell structures, are located by exposure to a photographic emulsion forming a pattern on the film corresponding to the location of the radioactive compounds within the cell. A technique in which radioactive molecules make their location known by exposing photographic films or emulsions.

autoregulation. The control of the transcription of a gene by its own gene product.

autosomal set. The non-sex chromosomes consisting of one from each homologous pair in a diploid species.

autosome. (1) Any chromosome except a sex chromosome. (2) Any chromosome that is not a sex chromosome.

autotroph. Organism that is able to utilize carbon dioxide as a carbon source.

autozygosity. Homozygosity in which the two alleles are identical by descent (i.e., they are copies of an ancestral gene).

auxotroph. An organism with specific nutritional requirements.

auxotrophic mutant. A mutant strain of microorganism that will proliferate only when the medium is supplemented with some specific substance not required by wild-type organisms.

Axolotl (Ambystoma mexicanum). Literally meaning water monster (in Aztec), axolotl is a salamander (amphibian) extensively used in the evolutionary immunogenetic studies.

axoneme. A bundle of microtubules and other proteins forming the core of each cilium or flagellum.

B

B cells. A major family of small lymphocytes that are responsible for antigen-specific humoral immunity as part of the adaptive immunity. Their antigen receptors are surface immunoglobulins (antibodies). They recognize peptides directly and secrete antibodies by differentiating into plasma cells. They also exist as long-lived memory cells.

B Factor. A fungal incompatibility factor. It operates in the Basidiomycetes species **Schizophyllum commune** (not to be confused with Factor B of the immune system).

B form DNA. A right-handed double-helical conformation of DNA normally seen in solution described by Watson and Crick. A second DNA conformation (A form) is seen in unhydrated DNA (fibres or crystals of oligonucleotides).

BAC. Bacterial Artificial Chromosome — a cloning vector capable of carrying between 100 and 300 kilobases of target sequence. They are propagated as a mini-chromosome in a bacterial host. The size of the typical BAC is ideal for use as an intermediate in large-scale genome sequencing projects. Entire genomes can be cloned into BAC libraries, and entire BAC clones can be shotgun-sequenced fairly rapidly.

bacillus thuringiensis. This bacteria is pathogenic to insects and the gene for its toxin is used to create transgenic plants with their own insecticide.

bacillus. A rod-shaped bacterium.

back mutation. The process that causes reversion. A change in a nucleotide pair in a mutant gene that restores the original sequence and hence the original phenotype.

backcross. The cross of an individual with one of its parents or an organism with the same genotype as a parent.

background extinction. The normal constant occurrence of extinctions. See **extinction files** in BBC Education.

bacterial lawn. A continuous cover of bacteria on the surface of a growth medium.

bacteriophage lambda. A virus which infects *E. coli*, and which is often used in molecular genetics experiments as a vector, or cloning vehicle. Recombinant phages can be made in which certain non-essential l DNA is removed and replaced with the DNA of interest. The phage can accommodate a DNA "insert" of about 15-20 kb. Replication of that virus will thus replicate the investigator's DNA. One would use phage l rather than a plasmid if the desired piece of DNA is rather large.

bacteriophage(phage). A virus that infects bacteria.

balanced lethal system. An arrangement of recessive lethal alleles that maintains a heterozygous chromosome combination. homozygotes for any lethal-bearing chromosome perish.

balanced lethal. Lethal mutations in different genes on the same pair of chromosomes that remain in repulsion because of close linkage or crossover suppression. In a closed population, only the trans-heterozygotes (l_1+ / + l_2) for the lethal mutations survive.

balanced polymorphism. (1) Stable genetic polymorphism maintained by natural selection. (2) The maintenance of two or more alleles in a population due to a selective advantage of the heterozygote.

balancing selection. Selection involving opposing forces in which selective advantages and disadvantages cancel each other out. Heterozygote advantage (or overdominant selection) is an example in which an allele selected against in the homozygous state is retained because of the superiority of heterozygotes. Other balanced states may occur including when: an allele is favored at one developmental stage and is selected against at another (antagonistic pleiotropy); an allele is favored in one sex and selected against in another (sexual antagonism); an allele is favored when it is rare and selected against when it is common (negative frequency dependent selection).

Balbiani ring. A large chromosome puff. The larger polytene chromosomal puffs. See chromosome puff.

band shift assay. See Gel shift assay.

barr body. (1) A densely staining mass that represents an inactivated X chromosome. Heterochromatic body found in the nuclei of normal females but absent in the nuclei of normal males. (2) Also called sex-chromatin body which represents the inactivated X chromosome in the nucleus of somatic mammalian cells. Normally only seen in female cells and not in male cells.

basal body. A short cylindrical array of microtubules and other proteins, found at the base of a eukaryotic cilium or flagellum, that organises the assembly of the axoneme (the bundle of microtubules and other proteins forming the core of each cilium or flagellum).

base analogue. A chemical whose molecular structure mimics that of a DNA base; because of the mimicry, the analogue may act as a mutagen.

base. A compound, usually containing nitrogen, that can accept a H^+. It is used to describe the non-sugar components of nucleotides (despite the basic nature of nucleotides, nucleic acids are acidic due to the phosphate atoms they contain). The five bases that form the nucleic acids are adenine (A), guanine (G), cytosine (C), tymine (T) and uracil (U).

Batesian mimicry. Form of mimicry in which an innocuous mimic species gains protection by resembling noxious or dangerous model species.

bead theory. The disproved hypothesis that genes are arranged on the chromosome like beads on a necklace, indivisible into smaller units of mutation and recombination.

beta-galactosidase. The enzyme that splits lactose into glucose and galactose (coded by a gene (lac z) in the lac operon of Escherichia coli).

beta-galactoside acetyltransferase. An enzyme that is involved in lactose metabolism and encoded by a gene (**lac a**) in the lac operon of Escherichia coli).

beta-galactoside permease. An enzyme involved in concentrating lactose in the cell (coded by a gene (lac y) in the lac operon of Escherichia coli)).

β-pleated sheet. A planar secondary structure element of proteins. It is created by hydrogen bonding between the backbone atoms in two different polypeptide chains or segments of a single folded chain.

bimodal distribution. A statistical distribution having two modes.

binary fission. (1) Mode of reproduction not involving any sex but division of a parent cell into two equally sized offspring. (2) The process in which a parent cell splits into two daughter cells of approximately equal size. Simple cell division in single-celled organisms.

binding site. A place on cellular DNA to which a protein (such as a transcription factor) can bind. Typically, binding sites might be found in the vicinity of genes, and would be involved in activating transcription of that gene (promoter elements), in enhancing the transcription of that gene (enhancer elements), or in reducing the transcription of that gene (silencers). NOTE that whether the protein in fact performs these functions may depend on some condition, such as the presence of a hormone, or the tissue in which the gene is being examined. Binding sites could also be involved in the regulation of chromosome structure or of DNA replication.

binomial expansion. The terms generated when a binomial expression [e.g. (p+q)] is raised to a particular power n [e.g. (p+q)^n].

binomial theorem. The theorem that gives the terms of the expansion of a binomial expression [e.g. (p+q)] raised to a particular power n [e.g. (p+q)^n].

biochemical genetics. The study of the relationships between genes and enzymes, specifically the role of genes in controlling the steps in biochemical pathways.

biolistic. A method (biological ballistics) of transfecting cells by bombarding them with microprojectiles coated with DNA.

biological species concept. Organisms are classified in the same species if they are potentially capable of interbreeding and producing fertile offspring.

biome. A grouping of plant ecosystems into a large distinct group occupying a major terrestrial region. They are created and maintained by climate. See examples of biomes.

biosphere. The geographical region of the Earth where life is found.

biparental zygote. A Chlamydomonas zygote that contains chloroplast DNA (cpDNA) from both parents, such cells generally are rare since normally cpDNA is inherited uniparentally from the mt+ mating type parent. These rare biparental zygotes allowed mapping of chloroplast genes by recombination.

bipedal. Two-footed posture and locomotion like a human standing upright on the hindlimbs.

bivalent. Structure, formed during prophase of meiosis 1, consisting of synapsed homologous chromosomes. Equivalent to a tetrad of chromatids.

blastoderm. (1) In an insect embryo, the layer of cells that completely surrounds an internal mass of yolk.

blastomere. One of the cells formed by division of the fertilized egg making up the blastula.

blastopore. The embryonic structure present during gastrulation and formed by invagination of the endoderm.

blastula. An early developmental stage of vertebrate embryos formed by cleavage of the fertilized egg, in which the embryo consists of a single layer of cells surrounding a cavity (the blastocoel).

blending inheritance. A discredited model of inheritance suggesting that the characteristics of an individual result from the smooth blending of fluid like influences from its parents.

blotting. A technique for detecting one RNA within a mixture of RNAs (a Northern blot) or one type of DNA within of a mixture of DNAs (a Southern blot). A blot can prove whether that one species of RNA or DNA is present, how much is there, and its approximate size. Basically, blotting involves gel electrophoresis, transfer to a blotting membrane (typically nitrocellulose or activated nylon), and incubating with a radioactive probe. Exposing the membrane to X-ray film produces darkening at a spot correlating with the position of the DNA or RNA of interest. The darker the spot, the more nucleic acid was present there.

blunt-end ligation. The ligating or attaching of blunt-ended pieces of DNA by T4 DNA ligase. Used in creating hybrid vectors.

Bony fish (class Osteichthyes). The vertebrate Class evolved after jawless and **cartilaginous fishes**. They have jaws, their skeleton is made up of bone and their body is covered with overlapping scales. Most familiar freshwater and sea water fishes belong to this group. The living fossil Coelacanth is a bony fish whose relatives (lobe-finned fishes) can be traced back to the Devonian geological period (363-409 Mya).

botryllus schlosseri. A clonal hermaphrodite (Phylum Chordata, Subphylum Tunicata, Class Ascidian=sea squirts). It grows fast, reproduces weekly, and thus, a good model for genetic studies of protochordates. It also has a well-characterized allorecognition system called Fu/HC whose functions are prevention of fusion with non-kin and selective fertilization by sperm bearing Fu/HC alleles different from that of the egg.

bottleneck. (1) A brief reduction in size of a population which usually leads to random genetic drift. (2) A drastic reduction in the population size followed by an expansion. This often results in altered gene pool as a result of genetic drift.

BP. Abbreviation for base pair(s). Double stranded DNA is usually measured in bp rather than nucleotides (nt).

brachydactyly. A human phenotype of unusually short digits, generally inherited as an autosomal dominant.

branch migration. 1. The process in which a crossover point between two DNA duplexes slides along the duplexes.

2. The process by which a single invading DNA strand extends its partial pairing with its complementary strand as it displaces the resident strand from a DNA duplex.

breakage and reunion. The general mode by which recombination occurs. DNA duplexes are broken and reunited in a crosswise fashion according to the Holliday model.

breakage-fusion-bridge cycle. Damage that happens to a dicentric chromosome during each cell cycle.

bridging cross. A cross made to transfer alleles between two sexually isolated species by first transferring the alleles to an intermediate species that is sexually compatible with both.

broad heritability (H^2). The proportion of total phenotypic variance at the population level that is contributed by genetic variance.

bud. A daughter cell formed by mitosis in yeast; one daughter cell retains the cell wall of the parent, and the other (the bud) forms a new cell wall.

buoyant density of DNA. A measure of the density of DNA determined by the equilibrium point reached by DNA after density gradient centrifugation.

buoyant density. A measure of the tendency of a substance to float in some other substance; large molecules are distinguished by their differing buoyant densities in some standard fluid. Measured by density-gradient ultracentrifugation.

Burkitt lymphoma. A cancer of the lymphatic system manifested by tumours in the jaw, often associated with a translocation bringing a specific oncogene next to a novel regulatory element.

C

c value. The amount of DNA comprising the haploid genome for a given species (picograms per cell; 2-3 pg in mammals). The C value paradox is the lack of correlation between the C values of species and their evolutionary complexity. For example, some amphibians have 30 times as much DNA as we have but not more complex than humans.

C. Cytosine or cytidine.

CAAT box. A highly conserved DNA sequence found about 75 bp 5' to the site of transcription in eukaryotic genes. Its specific (trans-acting) transcription factor is CTF-1 (NF-1) (see also TATA / Goldberg-Hogness box).

caenorhabditis elegans. A normally self-fertilizing hermaphrodite soil nematode whose developmental genetics has been extensively studied. It is no more than 1 mm long. Loss of an X chromosome by meiotic disjunction leads to the production of males. The genetic basis of apoptosis was first shown in C.elegans in 1986. It has five equally sized chromosomes and it is the first animal whose whole genome has been sequenced (in 1998). The 97 Mbp genome contains 19,000 genes on 6 chromosomes. About 74% of human genes have their homologues in the C.elegans genome.

callus. An undifferentiated clone of plant cells.

Cambrian. Cambria is the old name for Wales (UK) where the first skeletalized animal fossils were found. The first period in the Palaeozoic era marked by the occurrence of many forms of invertebrate life (540-500 Mya). The sudden appearance of the major animal phyla in the fossil record during the Cambrian period is called Cambrian explosion.

cAMP (cyclic adenosine monophosphate). A molecule that plays a key role in the regulation of various processes within the cell.

canalized character. A character whose phenotype is kept within narrow boundaries even in the presence of disturbing environments or mutations.

cancer. A syndrome that involves the uncontrolled and abnormal division of eukaryotic cells. An informal term for a diverse class of diseases marked by abnormal cell proliferation.

cancer-family syndrome. Pedigree in which unusually large numbers of blood relatives develop certain kinds of cancers.

CAP (catabolite activator protein). A protein, which binds cyclic AMP, whose presence is necessary for the activation of operons whose activity is repressed by the presence of glucose (catabolite repression) e.g. lac (lactose), ara (arabinose), gal (galactose) etc. cAMP-CAP complex binds to a site, the CAP binding site, which is centred at -61 to -62 from the transcription start site in the lac operon promoter.

cap site. (1) The initiation site of transcription in a eukaryotic gene. The initiation of translation of most eukaryotic mRNAs involves recognition of the cap followed by either the first downstream AUG or by a 5' proximal AUG with a consensus sequence surrounding it (like the bacterial Shine-Dalgarno or the viral Kozak sequence). Such a consensus sequence has not been recognized in eukaryotes yet. (2) Two usages: In eukaryotes, the cap site is the position in the gene at which transcription starts, and really should be called the "transcription initiation site". The first nucleotide is transcribed from this site to start the nascent RNA chain. That nucleotide becomes the 5' end of the chain, and thus the nucleotide to which the cap structure is attached (see "Cap"). In bacteria, the CAP site (note the capital letters) is a site on the DNA to which a protein factor (the Catabolite Activated Protein) binds.

cap. (1) A methylated guanine residue (GTP) which is added to the 5' end of eukaryotic mRNAs in a post-transcriptional reaction. It protects the mRNA against 5'-exonuclease. (2) A methylated guanine residue added in reverse polarity (i.e. 3'pMeG5'ppp5'NpNp3') to the 5' end of eukaryotic mRNA during transcription initiation. The cap binds a cap binding protein and acts as an initial binding site for ribosomes during translation. (3) All eukaryotes have at the 5' end of their messages a structure called a "cap", consisting of a 7-methylguanosine in 5'-5' triphosphate linkage with the first nucleotide of the mRNA. It is added post-transcriptionally, and is not encoded in the DNA.

capsid. The protein shell of a virus.

capsomere. Protein clusters making up discrete subunits of a viral protein shell.

carbon source. A nutrient (such as sugar) that provides carbon skeletons needed for synthesis of new organic molecules (anabolism).

carcinogen. An agent (chemical or radiation) that causes cancer.

carcinoma. tumour arising from epithelial tissue (e.g., glands; breast; skin; linings of the urogenital, intestinal and respiratory systems).

carrier. An individual who possesses a mutant allele but does not express it in the phenotype because of a dominant allelic partner; thus, an individual of genotype Aa is a carrier of a, if there is complete dominance of A over a.

cartilaginous fish (Class Chondrichthyes). The most primitive 'jawed' vertebrates evolved about 400 Mya. Their skeleton is composed of entirely cartilage. The Class includes the sharks, rays and skates (subclass Elasmobranchii) and the ratfish (subclass Holocephali). The earliest taxon which has both MHC class I and class II genes. The next step in the evolutionary ladder is the bony fish.

caspase. Cysteine-containing aspartic acid-specific proteases involved in the execution phase of apoptosis. Fas/Fas ligand system is one activator of caspase-dependent apoptotic cell death.

cassette mechanism. The mechanism by which homothallic yeast cells alternate mating types, involves two silent transposons (cassettes) and a region where these cassettes can be expressed (cassette player).

cassette model. A model to explain mating-type switching in yeast. Information for both a and alpha mating types is assumed to be present as silent cassettes; a copy of either type of cassette may be transposed to the mating-type locus, where it is played (transcribed).

CAT assay. An enzyme assay. CAT stands for chloramphenicol acetyl transferase, a bacterial enzyme which inactivates chloramphenicol by acetylating it. CAT assays are often performed to test the function of a promoter. The gene coding for CAT is linked onto a promoter (transcription control region) from another gene, and the construct is "transfected" into cultured cells. The amount of CAT enzyme produced is taken to indicate the transcriptional activity of the promoter (relative to other promoters which must be tested in parallel). It is easier to perform a CAT assay than it is to do a Northern blot, so CAT assays were a common method for testing the effects of sequence changes on promoter function. Largely supplanted by the reporter gene luciferase.

catabolite activator protein (CAP). A protein that when bound with cyclic AMP can attach to sites on sugar-metabolizing operons to enhance transcription of these operons.

catabolite repression. Repression (inactivation) of certain sugar-metabolizing operons (e.g., lac) in favour of glucose utilization when glucose is the predominant carbon source in the environment of the cell.

catalyst. A substance that increases the rate of a chemical reaction without itself being permanently changed.

catarrhini. One of the two divisions (suborder) of Primates containing the old world monkeys and apes (extinct and extant). The other division is Platyrrhini (new world monkeys).

cation. A positively charged ion (such as K^+, Na^+, NH4^+).

CCAAT box. (1) (CAT box, CAAT box, other variants) A sequence found in the 5' flanking region of certain genes which is necessary for efficient expression. A transcription factor (CCAAT-binding protein, CBP) binds to this site. (2) An invariant DNA sequence at about minus 70 base pairs from the origin of transcription in many eukaryotic promoters.

cDNA clone. "Complementary DNA"; a piece of DNA copied from an mRNA. The term "clone" indicates that this cDNA has been spliced into a plasmid or other vector in order to propagate it. A cDNA clone may contain DNA copies of such typical mRNA regions as coding sequence, 5'-untranslated region, 3' untranslated region or poly(A) tail. No introns will be present, nor any promoter sequences (or other 5' or 3' flanking regions). A "full-length" cDNA clone is one which contains all of the mRNA sequence from nucleotide #1 through to the poly(A) tail.

cDNA library. A pool of complementary DNA clones produced by cDNA cloning of total messenger RNA from a single source (cell type, tissue, embryo).

cDNA. See complementary DNA.

cell autonomous. A genetic trait in multicellular organisms in which only genotypically mutant cells exhibit the mutant phenotype. Conversely, a nonautonomous trait is one in which genotypically mutant cells cause other cells (regardless of their genotype) to exhibit a mutant phenotype.

cell cycle. The cycle of cell growth, replication of the genetic material and nuclear and cytoplasmic division.

cell division. The process by which two cells are formed from one. See meiosis and mitosis

cell fate. The ultimate differentiated state to which a cell has become committed.

cell lineage. A pedigree of cells related through asexual division.

cell-free system. A mixture of cytoplasmic and/or nuclear components from cells and used for in vitro protein synthesis or transcription or DNA replication or other purposes.

cellular blastoderm. The stage of the blastoderm, initially a syncytium, in insects after the nuclei have each been separately packaged into a cell membrane.

centimorgan (cm). A chromosome mapping unit. One centimorgan equals 1% recombinant offspring.

central dogma. The original postulate that genetic information can be transferred only from nucleic acid to nucleic acid and from nucleic acid to protein, that is from DNA to DNA from DNA to RNA and from RNA to protein (although information transfer from RNA to DNA was not excluded and is now known to occur [reverse transcription]). Transfer of genetic information from protein to nucleic acid never occurs.

centric fragment. A piece of chromosome containing a centromere.

centric fusion. Fusion of the long arms of two acrocentric chromosomes [13,14,15,21,22] into a single chromosome having lost the short arms at the same time. Most often occurs as 21/21, 13/14, and 14/21 translocations. Apart from being an important cause of uniparental disomy, it may cause trisomy 21 (Down's syndrome) in the offspring. Human chromosome 2 is a result of a centric fusion between two ancestral ape chromosomes (gorillas have 24 pairs of chromosomes).

centriole. A short cylindrical organelle, found in pairs arranged at right angles to each other at the centre of a microtubule organizing centre (MTOC) or centrosome, found in eukaryotes (except in higher plants). A centriole is similar in structure to the basal body found at the base of eukaryotic cilia and flagella and organises the axoneme, the bundle of microtubules and other proteins forming the core of each cilium or flagellum. The centrosome organizes formation of a spindle during mitosis or meiosis.

centromere marker. Locus located close to a centromere.

centromere. (1) A kinetochore; the constricted region of a nuclear chromosome, to which the spindle fibres attach during division. (2) Constricted region where sister chromatids are attached in mitotic chromosomes. The centromere is generally flanked by repetitive DNA sequences and it is late to replicate. The centromere is an A-T region of about 130 bp. It binds several proteins with high affinity to form the kinetochore which is the anchor for the mitotic spindle.

centromeric fission. Creation of two chromosomes from one by splitting the centromere.

centrosome. The primary microtubule organising center (MTOC) of animal cells, that divides prior to cell division each daughter MTOC acts as one pole of the spindle apparatus. The centrosome usually contains a pair of centrioles.

cetacea. An Order of marine mammals. The suborder Odontocetti includes dolphins, killer whales and toothed whales. Many of the great whales (such as the blue whale) belong to a different suborder (baleen whales or Mysticetti).

chaperone. (1) Any cellular protein that binds to an unfolded or partially folded target protein to prevent misfolding, aggregation, and/or degradation of it. Chaperones also facilitate the target protein's proper folding, translocation and assembly within cells, preventing inappropriate interactions with other proteins. (2) See molecular chaperone.

character difference. Alternative forms of the same attribute within a species.

character displacement. Forced evolution of dissimilar characters in related species where their ranges overlap. Elsewhere, where they exist on their own, their similarities are maintained.

character release. This is the opposite of character displacement. Two closely related species become more alike in regions where their ranges do not overlap than in regions where they do.

character. An attribute of individuals within a species for which heritable differences can be defined.

Chargaff's rule. Chargaff's observation that in the base composition of DNA the quantity of adenine equaled the quantity of thymine and the quantity of guanine equaled the quantity of cytosine.

Charon phage. Phage lambda derivative used as vector in DNA cloning.

chase. See pulse-chase experiment.

chi site. Sequence of DNA at which the RecBCD protein cleaves one of the strands during recombination.

chiasma (plural chiasmata). (1) A cross-shaped structure commonly observed between nonsister chromatids during meiosis; the site of crossing-over. X-shaped configuration seen in tetrads during the latter stages of prophase 1 of meiosis. They represent physical crossovers between DNA molecules. (2) The points of physical overlap of nonsister chromatids crossing-over in meiosis.

chi-like Sequence. An octamer nucleotide sequence (A/G - C/T - A/T - A/G - G - A/T - G - G) that creates a recombinational hotspot in the genome (originally discovered in coliphage lambda). MHC class I transmembrane domain length variation, frequent gene conversions and deletions in the MHC-linked 21-hydroxylase gene (CYP21), gene conversions within the MHC class II genes in mice and humans, many oncogene translocations (BCL2 for example) are attributed to chi-like sequences at the breakpoint region. It acts like a restriction site for recombinase.

chimera. See mosaic.

chimeric plasmid. Hybrid or genetically mixed plasmid used in DNA cloning.

chi-square distribution. The sampling distribution of the chi-square statistic. A family of curves whose shapes depend on degrees of freedom.

chi-square test. A statistical test used to determine the probability of obtaining the observed results by chance, under a specific hypothesis.

chlamydomonas. The unicellular green alga that is probably the closest living organism to the ancestor of green plants. It reproduces both asexually and sexually (two mating types). When reproduces sexually, the mitochondria are inherited from the (-) mating type and chloroplasts from the (+) mating type.

chloramphenicol acetyl transferase (CAT). The bacterial gene for chloramphenicol, CAT, is commonly used as a reporter gene for investigating physiological gene regulation. Beta-galactosidase and luciferase genes can also be used for the same purpose.

chloroplast. (1) A major component of a plastid in green plants and eukaryotic algae of any color. It is involved in photosynthesis. Prokaryotic photosynthetic organisms do not have chloroplasts. (2) The organelle that carries out photosynthesis and starch grain formation. A chlorophyll-containing organelle in plants that is the site of photosynthesis.

chordata. A major phylum in the Kingdom Animalia. A chordate is characterized by the presence of a dorsal notochord at some stage of development and a dorsal hollow nerve chord.

chromatid conversion. Gene conversion that is inferred from the existence of identical sister-spore pairs in a fungal octad that shows a non-Mendelian allele ratio.

chromatid interference. A situation in which the occurrence of a crossover between any two nonsister chromatids can be shown to affect the probability of those chromatids being involved in other crossovers in the same meiosis.

chromatid. (1) One of the two side by side replicas produced by chromosome replication in mitosis or meiosis. Subunit of a chromosome after replication and prior to anaphase of meiosis II or mitosis. At anaphase of meiosis II or mitosis when the centromeres divide and the sister chromatids separate each chromatid becomes a chromosome. (2) One of two copies of a replicated chromosome during mitosis. Together they are called sister chromatids. Each one becomes a daughter chromosome at anaphase of mitosis and at the second meiotic division.

chromatin. (1) The complex of DNA and associated histone and non-histone proteins that represents the normal state of genes in the nucleus. It exists in two forms: euchromatin can be transcribed, and heterochromatin is highly condensed and cannot be transcribed. (2) The substance of chromosomes; now known to include DNA, chromosomal proteins, and chromosomal RNA. The nucleoprotein material of the eukaryotic chromosome.

chromista. A major taxon in the Kingdom Protoctista (also called Heterokonta). They are mostly photosynthetic algae but distinct from the rest of the algae. They have genetic material derived from four different ancestors (purple bacteria, cyanobacteria, red alga and green alga). Link to **Chromista**.

chromocenter. The point at which polytene chromosomes appear to be attached together.

chromomere. A small beadlike structure visible on a chromosome during prophase of meiosis and mitosis. Dark regions of chromatin condensation in eukaryotic chromosomes at meiosis or mitosis.

chromosome aberration. Any type of change in the chromosome structure or number.

chromosome jumping. A technique of isolating clones from a genomic library that are not contiguous by skipping a region between known points on the chromosome. Done usually to bypass regions that are difficult or impossible to walk through or regions known not to be of interest.

chromosome loss. Failure of a chromosome to become incorporated into a daughter nucleus at cell division.

chromosome map. See linkage map.

chromosome marker. Any distinct and heritable feature of chromosome structure that can be used to follow (usually by microscopy) that chromosome or chromosome region in breeding experiments.

chromosome mutation. Any type of change in the chromosome structure or number.

chromosome painting. The use of fluorescent-tagged chromosome-specific dispersed repeat DNA sequences to visualize specific chromosomes or chromosome segments by in situ DNA hybridization and fluorescence microscopy.

chromosome puff. A swelling at a site along the length of a polytene chromosome; the site of active transcription. A diffuse uncoiled region in a polytene chromosome where transcription is actively taking place.

chromosome rearrangement. A chromosome mutation involving new juxtapositions of chromosome parts.

chromosome set. The group of different chromosomes that carries the basic set of genetic information for a particular species.

chromosome theory of inheritance. The theory that chromosomes are linear sequences of genes. The unifying theory stating that inheritance patterns may be generally explained by assuming that genes are located in specific sites on chromosomes.

chromosome walking. (1) A technique for cloning everything in the genome around a known piece of DNA (the starting probe). You screen a genomic library for all clones hybridizing with the probe, and then figure out which one extends furthest into the surrounding DNA. The most distal piece of this most distal clone is then used as a probe, so that ever more distal regions can be cloned. This has been used to move as much as 200 kb away from a given starting point (an immense undertaking). Typically used to "walk" from a starting point towards some nearby gene in order to clone that gene. Also used to obtain the remainder of a gene when you have isolated a part of it. (2) A technique which produces sets of overlapping DNA clones for studying segments of DNA larger than can be cloned individually. A method for the analysis of large regions of DNA, in which a each end of a large single cloned DNA fragment is used separately to screen recombinant DNA genome library for other clones containing neighbouring sequences.

chromosome. (1) A linear end-to-end arrangement of genes and other DNA, sometimes with associated protein and RNA. The form of the genetic material in viruses and cells. A circle of DNA in prokaryotes; a DNA or an RNA molecule in viruses; a linear nucleoprotein complex in eukaryotes. (2) Structure in a cell nucleus that carries the genes. Each chromosome consists of one very long strand of DNA, coiled and folded to produce a compact body. They become more compact and visible during metaphase of cell division. In interphase chromosomes, chromatin fibers are organized into 30 to 100 kb loops anchored in a supporting matrix within the nucleus. The length of each DNA molecule must be compressed about 8000-fold to generate the structure of a condensed metaphase chromosome.

ciliates. This group of protists are most like animals in their behavior and complexity. The Paramecium is the representative of the Ciliates.

circular permutation. Some phages with linear genomes (e.g., T2) are both terminally redundant and circularly permuted. If the genetic information is represented by

ABCDEFGH

then a circular permutation would generate molecules

ABCDEFGH
BCDEFGHA
CDEFGHAB
DEFGHABC
EFGHABCD

and so on.

cis conformation. In a heterozygote involving two mutant sites (a,b) within a gene or within a gene cluster, the arrangement ab/AB.

cis dominance. The ability of a gene to affect genes next to it on the same chromosome.

cis. Meaning on the near side of; refers to geometric configurations of atoms or mutations on the same molecule or chromosome.

cis-acting gene. A gene acting on or co-operating with another gene on the same chromosome (see trans-acting gene).

cis-dominant. mutations (e.g., of an operator) that alter the functioning of genes on that same piece of DNA.

cis-trans complementation test. A mating test to determine whether two different recessive mutations (a1;a2) on opposite chromosomes (trans, a1+/+a2) of a diploid or partial diploid will not complement (i.e. have a mutant phenotype) each other; but the same two recessive mutations on the same chromosome (cis, a1a2/++) in a diploid or partial diploid show a wild-type phenotype. A test for allelism. A test to determine whether two mutant sites are in the same functional unit or gene.

cistron. (1) A DNA segment coding for a specific polypeptide, and includes its own start and stop codons. When an mRNA encodes two or more proteins, it is called polycistronic. (2) Term coined by Benzer for the smallest genetic unit that does NOT show genetic complementation when two different mutations are in trans during a cis-trans complementation test; but shows wild-type phenotype when the same mutations are in cis. Synonymous with gene. Originally defined as a functional genetic unit within which two mutations cannot complement. Now equated with the term gene, as the region of DNA that encodes a single polypeptide (or functional RNA molecule such as tRNA or rRNA).

clade. All descendants of any given species. A single whole branch of a phylogeny.

cladogenesis. The evolutionary process whereby one species splits into two or more species. See anagenesis.

class switching. The process by which a IgM or IgD producing B lymphocyte switches to produce one of the secondary immunoglobulins (IgG, IgA or IgE) with the same antigen binding specificity.

class. A category of classification (taxon); a subdivision of subphylum. The classes in the Subphylum Vertebrata are: Pisces (Fishes), Amphibia, Reptilia, Avis (Birds) and Mammalia.

ClB method. A technique devised by Muller to rapidly screen fruit flies for recessive X chromosome lethal mutations. The ClB chromosome carries a recessive lethal (l) a dominant marker (B) and an inversion (crossover suppressor C).

clinal selection. Selection that changes gradually along a geographic gradient.

clone (noun). The term "clone" can refer either to a bacterium carrying a cloned DNA, or to the cloned DNA itself. If you receive a clone from a collaborator, you should first figure out if they send you DNA or bacteria. If it is DNA, your first job is to introduce it ("transform" it) into bacteria [see "Transformation (with respect to bacteria)"]. Occasionally, someone might send just the "insert", rather than the whole plasmid. "You assignment, Jim, if you decide to accept it", is to splice that DNA into a convenient vector, and only then can you transform it into bacteria.

clone (verb). To "clone" something is to produce copies of it. To clone a piece of DNA, one would insert it into some type of vector (say, a plasmid) and put the resultant construct into a host (usually a bacterium) so that the plasmid and insert replicate with the host. An individual bacterium is isolated and grown and the plasmid containing the "cloned" DNA is re-isolated from the bacteria, at which point there will be many millions of copies of the DNA - essentially an unlimited supply. Actually, an investigator wishing to clone some gene or cDNA rarely has that DNA in a purified form, so practically speaking, to "clone" something involves screening a cDNA or genomic library for the desired clone. See also "Probe" for a description of how one might start a cloning project, and "Screening" for how the probe in used. One can also clone more complex organisms, with considerable difficulty. The much-publicized Scottish research that resulted in the sheep 'Dolly' exemplifies this approach.

clone. (1) A group of genetically identical cells or individuals derived by asexual division from a common ancestor. (2) (colloquial), An individual formed by some asexual process so that it is genetically identical to its parent. (3) See also DNA clone.

cm(centimorgan). See map unit.

cnidaria. See Coelamata.

coadaptation. Interaction of genes at the genotypic level. Natural selection acts on the complex product of such interactions rather than on individual locus. The correlated variation and adaptation present in two mutually dependent organisms.

coalescence theory. The evolutionary theory that estimates the time for divergence from the last common ancestor.

coalescence. Growing into each other, uniting into one whole.

coccus. A spherical bacterium.

code dictionary. A listing of the 64 possible codons and their translational meanings (the corresponding amino acids).

coding sequence. The portion of a gene or an mRNA which actually codes for a protein. Introns are not coding sequences; nor are the 5' or 3' untranslated regions (or the flanking regions, for that matter - they are not even transcribed into mRNA). The coding sequence in a cDNA or mature mRNA includes everything from the AUG (or ATG) initiation codon through to the stop codon, inclusive.

coding strand. (1) An ambiguous term intended to refer to one specific strand in a double-stranded gene. See "Sense strand". (2) The DNA strand with the same sequence as the transcribed mRNA (given U in RNA and T in DNA) and containing the linear array of **codons** which interact with anticodons of tRNA during translation to give the primary sequence of a protein. Compare with anticoding strand.

codominance. Equal effect on the phenotype of two alleles of the same locus (as opposed to recessive and dominant).

codominance. The situation in which a heterozygote shows the phenotypic effects of both alleles equally, (e.g., blood group antigens).

codon bias. Although several codons code for a single amino acid, an organism may have a preferred codon for each amino acid. This is called codon bias.

codon preference. The idea that for amino acids with several codons one or a few are preferred and are used disproportionately. They would correspond with tRNAs that are abundant.

codon. (1) A section of DNA (three nucleotide pairs in length) or RNA (three nucleotides in length) that codes for a single amino acid. A sequence of three RNA or DNA nucleotides that specifies (codes for) either an amino acid or the termination of translation. (2) In an mRNA, a codon is a sequence of three nucleotides which codes for the incorporation of a specific amino acid into the growing protein. The sequence of codons in the mRNA unambiguously defines the primary structure of the final protein. Of course, the codons in the mRNA were also present in the genomic DNA, but the sequence may be interrupted by introns.

coefficient of coincidence. In a three point cross the number of observed double crossovers divided by the number expected based on the observed occurrence of single crossovers. The ratio of the observed number of double recombinants to the expected number.

coefficient of relatedness. $r = n(0.5)^L$ where n is the alternative routes between the related individuals along which a particular allele can be inherited; L is the number of meiosis or generation links.

coefficient of relationship (r). The proportion of alleles held in common by two related individuals.

coelacanth. Any member of a zoological family of largely extinct lobe-finned fish (like the living genus Latimeria).

coelomata (Cnidaria). A phylum in the Animal Kingdom. They are invertebrate aquatic animals showing radial or biradial symmetry. Examples are corals, sea anemones, jelly fish. See also Cnidaria in the Shape of Life (PBS).

co-evolution. Joint evolution of two unrelated species that have a close ecological relationship resulting in reciprocal adaptations as happens between host and parasite, and plant and insect.

cognate molecule. A relative descended from a common ancestor. Usually used to describe the corresponding partner in a receptor-ligand complex.

cohesive end. (1) A single-stranded end to a linear duplex DNA molecule which can hydrogen-bond with a complementary single-strand base sequence from the end of the same or another DNA molecule. (2) Also known as sticky end. Overhanging ends of a double-stranded DNA molecule that are capable of hybridizing with complementary ends.

cointegrate. A fusion of two elements. An intermediate structure in replicative transposition. The product of the fusion of two circular elements to form a single, larger circle.

col plasmid. Plasmid that produces an antibiotic (colicin) used by the host to kill other strains of bacteria.

colicinogenic factor. See col plasmid.

colinearity. The one to one linear correspondence between the order of codons in a coding sequence and the order of amino acids in the protein encoded. A linear map of mutation sites within a gene corresponds to the linear location of amino acid substitutions within the polypeptide encoded by that gene.

colony. A visible clone of cells.

common ancestry. The state of two individuals when they are blood relatives. When two parents have a common ancestor their offspring will be inbred.

compartment. The existence of boundaries within the organism beyond which a specific clone of cells will never extend during development.

competence factor. A surface protein that binds extracellular DNA and enables the cell to be transformed.

competent. Able to take up exogenous DNA and thereby be transformed.

complementarity. The correspondence of DNA bases in the double helix such that adenine in one strand is opposite thymine in the other strand and cytosine in one strand is opposite guanine in the other. This relationship explains Chargaff's rule.

complementary (copy) DNA (cDNA). Single-stranded DNA produced from an RNA template (usually mRNA) by reverse transcriptase in vitro. It lacks the introns present in corresponding genomic DNA. It is most commonly made to use in PCR to amplify RNA (RT-PCR).

complementary DNA (cDNA). Synthetic DNA reverse transcribed from a specific RNA through the action of the enzyme reverse transcriptase. DNA synthesized by reverse transcriptase using RNA as a template.

complementary RNA (cRNA). Synthetic RNA produced by transcription from a specific DNA single stranded template.

complementation group. A group of mutant genes which do not complement each other. A cistron (determined by the cis-trans complementation test).

complementation test. A mating test to determine whether two different recessive mutations (a1;a2) on opposite chromosomes (trans, a1+/+a2) of a diploid or partial diploid will not complement (i.e. have a mutant phenotype) each other; but the same two recessive mutations on the same chromosome (cis, a1a2/++) in a diploid or partial diploid show a wild-type phenotype; a test for allelism. A test to determine whether two mutant sites are in the same functional unit or gene.

complementation. The production of a wild-type phenotype when two different mutations are combined in a diploid or a heterokaryon. The production of the wild-type phenotype by a cell or an organism that contains two mutant genes. If complementation occurs the mutations are almost certainly non-allelic (i.e. in different genes).

complete linkage. The state in which two loci are so close together that alleles of these loci are virtually never separated by crossing over.

complete medium. A culture medium that is enriched to contain all of the growth requirements of a strain of organisms.

component of fitness. A particular aspect in the life cycle of an organism upon which natural selection acts.

composite transposon. A transposon constructed of two IS (insertion sequence) elements flanking a control region that frequently contains host genes.

compound heterozygote. An individual who is affected with an autosomal recessive disorder having two different mutations in homologous alleles.

concordance. The amount of similarity in phenotype between individuals.

conditional mutation. A mutation that has the wild-type phenotype under certain (permissive) environmental conditions and a mutant phenotype under other (restrictive) conditions.

conditional-lethal mutation. A mutation that is lethal under one condition but not lethal under another condition.

confidence limits. A statistical term for a pair of numbers that predict the range of values within which a particular parameter lies for a given level of confidence (probability).

congenic. Animals which have been bred to be genetically identical except for a single gene locus. This is achieved by superimposing the locus of interest on the genetic background of another by first crossing two inbred lines followed by extensive (about 20 generations) backcrossing hybrids to one parental line (the background strain) while selecting for the alleles of the locus of interest of the other. The result is an inbred strain uniquely identified by a difference at a single locus.

conifers. Cone bearing trees. A Class of the Gymnospermae which includes needle-leaved trees such as pines and cypresses. Their flowers are in cones, and male and flower cones are separate. The oldest (britlecone pine) and the largest (sequoia) extant organisms belong to this Class. Their unique feature is the inheritance of cytoplasmic DNA (chloroplasts) via pollens.

conjugation tube. See pilus.

conjugation. A process whereby two cells come in contact and exchange genetic material. In prokaryotes the transfer is a one-way process. The union of two bacterial cells, during which chromosomal material is transferred from the donor to the recipient cell. Conjugation in Protozoa is a two-way process, genetic material is passed between each conjugant.

conjugation. In unicellular organisms, temporary cell contact between complementary genders and exchange of genetic material, as in the ciliate Paramecium aurelia, or one-way transfer of genes as in bacteria (see F Factor) and green alga Spirogyra.

consanguineous matings. Matings between two individuals who share a common ancestor in the preceding two or three generations.

consanguineous. Meaning between blood relatives; usually refers to inbreeding or incestuous matings.

consensus sequence. (1) A 'nominal' sequence inferred from multiple, imperfect examples. Multiple lanes of shotgun sequence can me merged to show a consensus sequence. The optimal sequence of nucleotides recognized by some factor. A DNA binding site for a protein may vary substantially, but one can infer the consensus sequence for the binding site by comparing numerous examples. For example, the (fictitious) transcription factor ZQ1 usually binds to the sequences AAAGTT, AAGGTT or AAGATT. The consensus sequence for that factor is said to be AARRTT (where R is any purine, i.e. A or G). ZQ1 may also be able to weakly bind to ACAGTT (which differs by one base from the consensus). (2) A sequence of nucleotides or amino acids in common between regions of homology in different but related DNA or RNA or protein sequences.

consensus sequence. The nucleotides or amino acids most commonly found at each positions of the sequences of related molecules.

conservative replication. A postulated mode of DNA replication in which an intact double helix acts as a template for a new double helix; known to be incorrect. A disproved model of DNA synthesis suggesting that one-half of the daughter DNA molecules should have both strands composed of newly polymerized nucleotides.

conserved sequence. An invariant sequence found in different DNA or RNA or protein sequences.

constant region. A region of an antibody molecule that is nearly identical with the corresponding regions of antibodies of different specificities.

constitutive heterochromatin. Heterochromatin that surrounds the centromere. Specific regions of heterochromatin always present and in both homologs of a chromosome. See satellite DNA.

constitutive mutation. A mutation which causes transcription to be no longer under regulatory control.

constitutive. Always expressed in an unregulated fashion (when referring to gene control).

contig. Several uses, all nouns. The term comes from a shortening of the word 'contiguous'. A 'contig' may refer to a map showing placement of a set of clones that completely, contiguously cover some segment of DNA in which you are interested. Also called the 'minimal tiling path'. More often, the term 'contig' is used to refer to the final product of a shotgun sequencing project. When individual lanes of sequence information are merged to infer the sequence of the larger DNA piece, the product consensus sequence is called a 'contig'.

continuous replication. The uninterrupted replication of DNA in the 5' to 3' direction using a 3' to 5' template.

continuous variation. Variation measured on a continuum rather than in discrete units or categories (e.g., height in human beings).

controlling element. A term used by maize geneticists to indicate a mobile genetic element capable of producing an unstable mutant target gene; two types exist, the regulator and the receptor elements.

convergent evolution. Evolution of two or more different lineages towards similar morphology due to similar adaptive pressures. Examples of convergence are: fins or fin-like structures in fish, cuttlefish and whales; extreme similarity in alarm calls by five small birds; endothermy in dogs and ducks, wings of butterflies and birds.

copper fist. Configuration of a DNA-binding protein that resembles a fist closed around a penny. In this case the penny is copper ions; the knuckles of the fist of the yeast **ACE1** protein interact with the promoter of the metallothionein gene, enhancing its transcription.

copy-choice hypothesis. An incorrect hypothesis that stated that recombination resulted from the switching of the DNA-replicating enzyme from one DNA homologue to the other.

copy-choice model. A model of the mechanism for crossing over, suggesting that crossing over occurs during chromosome division and can occur only between two supposedly new nonsister chromatids; the experimental evidence does not support this model.

corepressor. The metabolite that when bound to the repressor (of a repressible operon) forms a functional unit that can bind to its operator and block transcription.

correction. The production (possibly by excision and repair) of a properly paired nucleotide pair from a sequence of hybrid DNA that contains a mismatched base pair. See mismatch repair

correlation coefficient. A statistic that gives a measure of how closely two variables are related. A statistical measure of the extent to which variations in one variable are related to variations in another.

cosegregation. The tendency for closely linked genes and genetic markers to segregate (be inherited) together.

cosmid. (1) A hybrid plasmid that contains cos sites at each end. Cos sites are recognized during head filling of lambda phages. Cosmids are useful for cloning large segments of foreign DNA (up to 50 kb). (2) A type of vector used for cloning 35-45 kb of DNA. These are plasmids carrying a phage l cos site (which allows packaging into l capsids), an origin of replication and an antibiotic resistance gene. A plasmid of 40 kb is very difficult to put into bacteria, but can replicate once there. Cosmids, however, have a cos site, and thus can be packaged into l phage heads (a reaction which can be performed *in vitro*) to allow efficient introduction into bacteria (you'll have to look up the cos site elsewhere).

cot value (cot1/2). The product of Co (the original concentration of denatured DNA) and t (time in seconds), giving a useful index of DNA renaturation. Cot1/2 is the value when 50% renaturation has occurred which can be used to estimate the length of unique DNA in a sample.

cotransduction. The simultaneous transduction of two or more genes. The simultaneous transduction of two bacterial marker genes.

cotransformation. The simultaneous transformation of two bacterial marker genes.

coupling (cis-arrangement). The condition in which a double heterozygote has received two linked mutations from one parent and their wild-type alleles from the other parent, e.g., *a b* / + + (as opposed to *a* + / + *b*; see also **repulsion**).

coupling conformation. Linked heterozygous gene pairs in the arrangement, AB/ab.

coupling. Arrangement of wild-type and mutant alleles at two linked loci in which both mutants are on the same chromosome and both wild-type alleles on the homologue (ab/AB). See repulsion

covariance. A statistical measure used in computing the correlation coefficient between two variables; the covariance is the mean of (x- x(bar))(y-y(bar)) over all pairs of values for the variables x and y, where x(bar) is the mean of the x values and y(bar) is the mean of the y values. A statistical value measuring the simultaneous deviations of x and y variables from their means.

cpDNA. Chloroplast DNA.

CpG island. Repetitive CpG doublets creating a region of DNA greater than 200 bp in length with a G+C content of more than 0.5 and an observed/expected presence of CpG more than 0.6. Usually associated with transcription-initiation regions of (housekeeping) genes transcribed at low rates that do not contain a TATA box. The CpG-rich stretch of 20-50 nucleotides occurs within the first 100-200 bases upstream of the start site region (where **promoter-proximal elements** reside). A trans-acting transcription factor called SP1 recognizes the CpG islands (see also **Htf islands**). In vertebrates, many of the nontranscribed genes (and the genes on the inactivated X chromosome) have a 5-methyl group on the C residue in CpG dinucleotides in transcription-control regions. On the other hand, many genes with restricted expression patterns have (methylated) CpG islands located downstream of transcription initiation does not block elongation of the transcript (see also **Methylation paradox**).

cretaceous. The youngest of the three geologic periods of the Mesozoic era (145 to 65 Mya).

cri-du-chat syndrome. A non-lethal human condition in infants caused by deletion of part of one homologue of chromosome 5.

crisscross inheritance. Transmission of a gene from male parent to female child to male grandchild for example, X-linked inheritance.

cRNA. See complementary RNA.

cross. The deliberate mating of two parental types of organisms in genetic analysis.

crossbreed. Fertilization between separate individuals.

cross-fertilization. See crossbreed.

crossing over. A process in which homologous chromosomes exchange parts normally reciprocally but sometimes unequally. The exchange of corresponding chromosome parts between homologues by breakage and reunion of DNA molecules normally during prophase I of meiosis but also occasionally during mitosis.

crossing-over (recombination). The exchange of genetic material between non-sister chromatids of homologous chromosomes (i.e., between maternal and paternal chromosomes) during meiosis. This results in a new and unique combination of genes on the daughter chromosome which will be passed on to the offspring (if that particular gamete is involved in fertilization).

crossover suppression. Reduction of crossing over within an inversion loop in inversion heterozygotes due to physical constraints during synapsis. Crossing over within an inversion loop, when it does occur, leads to defective (deleted and duplicated) crossover chromosomes and mortality of zygotes carrying them.

crossover suppressor. See Crossover suppression

crossover. See crossing over and chiasmata.

cruciform configuration. A region of DNA having a sequence at one end repeated but inverted at the other end, so that each strand may pair with itself to form a helix extending sideways from the main helix.

crustacean. Any of a zoological class (Crustacea) that have a chitinous and/or calcareous exoskeleton (lobsters, shrimps, crabs).

cryptic coloration. Coloration that allows an organism to match its background and hence become less vulnerable to predation or recognition by prey.

cryptic female choice. Besides precopulatory female sexual selection, there are also postcopulatory selection processes going on in the female reproductive system. This less appreciated mechanism is the basis for differential fertilization which includes sperm selection as opposed to pollen selection in plants. This should not be confused with sperm competition/pollen tube competition.

culture. Tissue or cells multiplying by asexual division, grown for experimentation.

cyanobacteria. Unicellular, photosynthetic (photo-autotroph) prokaryote (in the Kingdom Monera). Formerly known as blue-green algae. It contains chlorophyll *a* but not chloroplast. They reproduce by fission and never sexually.

cyclic adenosine monophosphate. See cAMP.

cyclic AMP (cAMP). A form of AMP (adenosine monophosphate) used frequently as a second messenger in eukaryotics and in catabolite repression in prokaryotes.

Cys. Cysteine (an amino acid).

cystic fibrosis (CF). A potentially lethal human disease of secretory cells, showing excess lung mucus secretion, and inherited as an autosomal recessive on chromosome 7. CF is caused by mutations in a gene encoding the cystic fibrosis membrane conductance regulator, a transmembrane protein involved in ion transport.

cytidine. The nucleoside containing cytosine as its base.

cytochrome. A class of protein, found in mitochondrial membranes, whose main function is oxidative phosphorylation of ADP to form ATP.

cytogenetics. The cytological approach to genetics, mainly involving microscopic studies of chromosomes.

cytohet. A cell containing two genetically distinct types of a specific organelle.

cytokinesis. The division of the cytoplasm of a cell into two daughter cells. See karyokinesis.

cytoplasm. The material between the nuclear and cell membranes; includes fluid (cytosol) organelles, and various membranes.

cytoplasmic inheritance. Inheritance via genes found in cytoplasmic organelles. Extra-chromosomal inheritance controlled by non-nuclear genomes.

cytosine. A pyrimidine base that pairs with guanine. See pyrimidine.

cytosol. The fluid portion of the cytoplasm, outside the organelles.

cytotoxic T lymphocyte (Tc). T lymphocyte responsible for attacking cancerous host cells or cells infected with an invading bacterium or virus.

D

Darwinian fitness. The relative probability of survival and reproduction for a genotype. see fitness.

dauermodification. The persistence for several generations of an environmentally induced trait.

de novo. Literally, 'from new' as opposed to inherited.

deficiency. The absence of part of the normal genome or chromosome set. See deletion.

degeneracy. A feature of the genetic code that more than one nucleotide triplet codes for the same amino acid. The same applies to the termination signal which is encoded by three different stop codons.

degenerate code. A code in which several code words have the same meaning. The genetic code is degenerate because there are many instances in which different codons specify the same amino acid. A genetic code in which some amino acids may each be encoded by more than one codon.

degrees of freedom. An estimate of the number of independent categories in a particular statistical test or experiment.

deletion chromosome. A chromosome containing a deletion.

deletion. Loss of a DNA (chromosome) segment from a chromosome. Deletions are recognised genetically by: 1. absence of reverse mutation. 2. presence of a deletion loop at meiosis visualized cytologically. 3. revealing of recessive lethals. 4. pseudodominance.

deme. A locally interbreeding population.

denaturation map. A map of a stretch of DNA showing the locations of local denaturation loops, which correspond to regions of high AT content.

denaturation. (1) The separation of the two strands of a DNA double helix, or (2) the severe disruption of the hydrogen bonded structure of any complex molecule without breaking the covalent bonds of its chains. (3) Reversible disruption of hydrogen bonds between nucleotides converting a double-stranded DNA molecule to single-stranded molecules. Heating or strong alkali treatment result in denaturation of DNA.

denatured. Loss of natural configuration (of a molecule) through heat or other treatment. Fully denatured DNA is single-stranded.

denominator element. Gene on an autosome of Drosophila that helps regulate the sex switch (sxl) to the off condition (maleness). Refers to the denominator of the X/A (number of X chromosomes/autosomes) genic balance equation that determines sex in insects.

density-gradient centrifugation. A method of separating macromolecules by their (1) differential rate of sedimentation in a centrifugal gradient (2) differential bouyancy in a density gradient.

deoxyribonuclease. See DNase.

deoxyribonucleic acid. See DNA.

depauperate fauna. A fauna, especially common on islands, lacking many species found in similar habitats elsewhere.

derepressed. The condition of an operon that is transcribing because repressor control has been lifted. May apply more generally to any gene being transcribed.

determinant. A spatially localized molecule that causes cells to adopt a particular fate or set of related fates.

determination. The process of commitment of cells to particular fates.

deterministic. Referring to events that have no random or probabilistic aspects but proceed in a fixed predictable fashion.

development. The process whereby a single cell becomes a differentiated organism. The process of orderly change that an individual goes through in the formation of structure.

Devonian Period. A geologic period in the Phanerozoic Eon (410 to 360 Mya).

diakinesis. The final stage of prophase 1 of meiosis when chiasmata terminalize.

dicentric chromosome. (1) A chromosome with two centromeres. (2) One chromosome having two centromeres.

dictyotene. A prolonged diplonema of primary oocytes that can last many years.

dideoxy method. A method of DNA sequencing that uses chain-terminating (dideoxy) nucleotides.

differentiation. The changes in cell shape and physiology associated with the production of the final cell types of a particular organ or tissue.

dihybrid cross. A cross between two individuals identically heterozygous at two loci for example, AaBb/AaBb.

dihybrid. An organism heterozygous at two loci.

dikaryotic. A cell that contains two separate haploid nuclei (n+n) which is different from being haploid (n) or diploid (2n). Naturally seen in fungal heterokaryons. Dikaryosis is a significant genetic peculiarity of the fungi.

dimerization. The chemical union of two identical molecules.

dimorphism. A polymorphism involving only two forms.

dioecious plant. A plant species in which male and female organs appear on separate individuals.

diploblast. A lower invertebrate such as jelly fish that are composed of two tissue layers (ectoderm and endoderm) and lacking the third layer (mesoderm) present in higher invertebrates and vertebrates.

diploid number (2n). The full complement of chromosomes in a somatic cell (or a sex cell before meiosis). In humans, the diploid number is 46.

diploid. The state of having each chromosome in two copies per nucleus or cell. A cell having two chromosome sets, or an individual having two chromosome sets in each of its cells.

diplonema (diplotene stage). The stage of prophase of meiosis I in which chromatids appear to repel each other.

directed mutagenesis. Altering some specific part of a cloned gene and reintroducing the modified gene back into the organism.

directional selection. A type of selection that removes individuals from one end of a phenotypic distribution and thus causes a shift in the distribution. Selection that changes the frequency of an allele in a constant direction, either toward or away from fixation for that allele.

disassortative mating. The mating of two individuals with dissimilar phenotypes.

discontinuous replication. Replication of DNA in short 5' to 3' segments using the 5' to 3' strand as a template while going backward away from the replication fork.

discontinuous variation. Variation that falls into discrete categories (e.g., the colour of garden peas).

discrete generations. Generations that have no overlapping reproduction. All reproduction takes place between individuals in the same generation.

disjunction. Separation of homologous chromosomes during anaphase of mitotic or meiotic divisions (see also nondisjunction).

dispersive replication. Disproved model of DNA synthesis suggesting more-or-less random interspersion of parental and new segments in daughter DNA molecules. A postulated mode of DNA replication combining aspects of conservative and semiconservative replication; known to be incorrect.

disposable soma theory. A theory on the evolution of ageing and death suggesting that organisms derive little benefit from investing resources in increasing their lifespan beyond a certain point. It originated from the economic phenomenon that manufacturers invest minimum in durability.

disruptive selection. (1) A type of selection that removes individuals from the centre of a phenotypic distribution and thus causes the distribution to become bimodal. (2) Selection against the middle range of variation causing an increase in the frequency of a trait showing the extreme ranges of its variation. Disruptive selection might cause one species to evolve into two.

distribution function. A graph of some precise quantitative measure of a character against its frequency of occurrence.

distribution. See statistical distribution.

disulfide bond (-S-S-). A covalent linkage between two cysteine residues in different parts of a protein or between two different proteins. Insulin (a small protein having two polypeptide chains) and immunoglobulin molecules, for example, have interchain and intrachain disulfide bonds. HLA molecules also have disulfide bonds. The C282Y mutation removes one of the disulfide bonds in the HLA class I-like HFE protein and abolishes its surface expression.

divergent evolution. A kind of evolutionary change that results in increasing morphological difference between initially more similar lineages.

d-loop. Configuration found during DNA replication of chloroplast and mitochondrial chromosomes wherein the origin of replication is different on the two strands. The first structure formed is a displacement loop or D-loop.

dn/ds ratio. In molecular phylogenetic studies, the ratio of the number of non-synonymous nucleotide substitutions to the number of synonymous nucleotide substitutions. In the case of functionally important (or otherwise constrained) genes, ds is expected to exceed dn (dn/ds <1). Because most amino acid changes will disrupt protein structure and those non-synonymous substitutions (dn) causing them will not be maintained. In a non-functional pseudogene, there will be no discrimination between them and equal numbers of dn and ds are expected (dn/ds=1). When natural selection is acting to favor changes at the amino acid level, it is predicted that dn will exceed ds, hence a high dn/ds ratio. In classical MHC loci, in the peptide binding regions (allele-specific sequences) because of heterozygote advantage/frequency-dependent selection, there is always a high dn/ds ratio (>1) whereas in the remainder of the gene dn/ds <1 (due to functional constrains). This suggests balancing selection is acting on peptide binding regions.

DNA (deoxyribonucleic acid). (1) An antiparallel double helix of nucleotides (having deoxyribose as their sugars) linked by phosphodiester (sugar-phosphate) bonds to adjacent nucleotides in the same chain and by hydrogen bonds to complementary nucleotides in the opposite chain. The fundamental substance of which genes are composed. (2) The large double-stranded molecule carrying the genetic code. It consists of four bases (adenine, guanine, cytosine and thymine), phosphate and ribose.

DNA binding motif. Common sites on different proteins which facilitate their binding to DNA. Examples are leucine zipper and zinc finger proteins. Any such protein is called DNA-binding protein.

DNA clone. A section of DNA that has been inserted into a vector molecule, such as a plasmid or a phage chromosome, and then replicated to form many identical copies. See also clone

DNA cloning. See gene cloning.

DNA fingerprint. The largely individual-specific autoradiographic banding pattern produced when DNA is digested with a restriction endonuclease that cuts outside a family of VNTRs, and a Southern blot of the electrophoretic gel is probed with a VNTR-specific probe.

DNA glycosylase. Endonuclease initiating excision repair at various damaged or improper bases in DNA.

DNA gyrase. A topoisomerase that relieves supercoiling in DNA by creating a transient break in the double helix.

DNA ligase. An enzyme that closes nicks or discontinuities in one strand of double-stranded DNA by creating an ester bond between adjacent 3' OH and 5' PO4 ends on the same strand.

DNA marker. Any unique DNA sequence which can be used in hybridization, PCR or restriction mapping experiments to identify that sequence. See expressed sequence tag and sequence tagged site.

DNA Polymerase. (1) A group of enzymes mainly involved in copying a single-stranded DNA molecule to make its complementary strand. Eukaryotic DNA polymerases participate in chromosomal replication, repair, crossing-over and mitochondrial replication. To initiate replication, DNA polymerases require a priming RNA molecule. They extend the DNA using deoxyribonucleotide triphosphates (dNTP) as substrates and releasing pyrophosphates. The dNMPs are added to the 3' OH end of the growing strand (thus, DNA replication proceeds from 5' to 3' end). (2) An enzyme that can synthesize new DNA strands using a DNA template; several such enzymes exist. One of several classes of enzymes that polymerize DNA nucleotides using single or double-stranded DNA as a template.

DNA repair. Restoration of the correct nucleotide sequence of a DNA molecule that has acquired a mutation or modification. It includes proofreading by **DNA polymerase** (see **helicase**).

DNA-DNA hybridization. When DNA is heated to denaturation temperatures to form single strands and then cooled double helices will re-form (renaturation) at regions of sequence complementarity. This technique is useful for determining sequence similarity among DNAs of different origin and the amount of sequence repetition within one DNA.

DNA-RNA hybridization. When a mixture of DNA and RNA is heated to denaturation temperatures to form single strands and then cooled RNA can hybridize (form a double helix) with DNA that has a complementary nucleotide sequence. Useful for determining relationships between DNAs and RNAs.

DNase. (1) Deoxyribonuclease, a class of enzymes which digest DNA. The most common is DNase I, an endonuclease which digests both single and double-stranded DNA. (2) An enzyme that degrades DNA to nucleotides.

docking protein. Responsible for attaching (docking) a ribosome to a membrane by interacting with a signal particle attached to a ribosome destined to be membrane bound.

domain. Region of a protein with a distinct tertiary structure and characteristic activity (for example, the membrane distal and membrane proximal domains of an MHC molecule).

dominance variance. Genetic variance at a single locus attributable to dominance of one allele over another.

dominance. The property possessed by some alleles of determining the phenotype for any particular gene by masking the effects of the other allele (when heterozygous). Thus, homozygosity or heterozygosity for the dominant allele result in the same genotype in complete dominance (if red is dominant over white, the petals of a flower heterozygous for red and white would be red). Incomplete dominance appears as a blend of the phenotypes corresponding to the two alleles (like pink petals as opposed to red or white).

In co-dominance, both alleles equally contribute to the phenotype (red and white petals occur together).

dominant allele. An allele that expresses its phenotypic effect even when heterozygous with a recessive allele; thus if **A** is dominant over a, then AA and Aa have the same phenotype.

dominant allele. An allele that masks an alternative allele when both are present (in heterozygous form) in an organisms (see recessive).

dominant phenotype. The phenotype of a genotype containing the dominant allele; the parental phenotype that is expressed in a heterozygote.

dominant. An allele that determines phenotype even when heterozygous. Also the trait controlled by that allele.

dominant-negative mutation. A (heterozygous) dominant mutation on one allele blocking the activity of wild-type protein still encoded by the normal allele causing a loss-of-function phenotype. The phenotype is indistinguishable from that of homozygous dominant mutation. p53 mutations may act as dominant-negative (see also haploinsufficiency).

dosage compensation. (1) A mechanism by which species with sex chromosomes ensure that the homogametic sex does not have too much or the heterogametic sex too little activity of loci on the homogametic sex chromosome. The process, in organisms using a chromosomal sex determination mechanism (such as XX versus XY), that allows standard structural genes on the homogametic sex chromosome to be expressed at the same levels in females and males, regardless of the number of homogametic sex chromosomes. In mammals, dosage compensation operates by maintaining only a single active X chromosome in each cell (see Lyon hypothesis), while in Drosophila it operates by hyperactivating the single male X chromosome. (2) The phenomenon in women, who have two copies of genes on the X chromosome, of having the same level of the products of those genes as males (who have a single X chromosome). This is due to the process of inactivation of one of the X chromosomes in females (Lyonization). See Lyon's hypothesis.

dose. See gene dose.

dot blot. A technique for measuring the amount of one specific DNA or RNA in a complex mixture. The samples are spotted onto a hybridization membrane (such as nitrocellulose or activated nylon, etc.), fixed and hybridized with a radioactive probe. The extent of labeling (as determined by autoradiography and densitometry) is proportional to the concentration of the target molecule in the sample. Standards provide a means of calibrating the results.

dot blotting. A technique of blotting cloned DNA without prior restriction digestion and electrophoresis (See Southern blotting). Autoradiography reveals dots indicating probe hybridization.

double crossover. Two crossovers occurring in a chromosomal region under study.

double digest. The product formed when two different restriction endonucleases act on the same sample of DNA.

double fertilization. The union of one male gametophyte nucleus with the ovum nucleus and another male gametophyte nucleus with the polar nuclei to form the endosperm in seed plants. See also pollen grain.

double helix. The normal structural configuration of DNA consisting of two helices winding about the same axis. The structure of DNA first proposed by Watson and Crick with two interlocking helices joined by hydrogen bonds between paired bases.

double heterozygote. An individual who is heterozygous at two loci under investigation.

double infection. Infection of a bacterium with two genetically different phages.

double reduction. The condition in polyploids in which a heterozygous individual produces homozygous gametes.

double stranded RNA (dsRNA). In eukaryotes, it is an accidental byproduct of transcriptional process. It may occur as the genome of certain viruses (such as reovirus) or may be produced during viral replication as a general marker for viral infection. It is believed that dsRNA is the toxic substance responsible for general symptoms of viral infection as it induces cytokine production. dsRNA is the major activator of the PKR enzyme which is the major agent of anti-viral innate immunity.

doublesex. Locus (dsx) in Drosophila in which some mutations convert both males and females into intersexes. Encodes male and female specific proteins produced by differential splicing of the primary transcript in each sex.

Down syndrome. An abnormal human phenotype, including mental retardation, due to a· trisomy of chromosome 2l; more common in babies born to older mothers.

downstream. (1) A convention used to describe features of a DNA sequence, gene or mRNA related to the position and direction (5' to 3') of transcription by RNA polymerase or translation by the ribosome. Downstream (or 3' to) is in the direction of transcription (or translation) whereas upstream (5' to) is in the direction from which the polymerase (or ribosome) has come. Conventionally DNA sequences, gene maps and RNA sequences are drawn with transcription (or translation) from left to right and so downstream is towards the right. (2) See "Upstream/ Downstream". (3) The direction which RNA polymerase moves during transcription (5' to 3') and ribosomes moves during translation. By convention, the +1 position of a gene is the first transcribed nucleotide, nucleotides downstream from the +1 position are designated +2, +3, etc.

drift evolution. A high rate of immunologically significant mutations in certain viruses. This results in drifting away from recognition by the immune system by antigenic change. Influenza virus, HIV and HCV constantly change their antigenic structure through drift evolution.

drift. See random genetic drift.

drosophila melanogester. Common fruit fly. It contributed heavily to the study of genetics because of its ease and speed to breed. It contains only four pairs of chromosomes.

Duchenne muscular dystrophy. A lethal muscle disease in humans caused by mutation in a huge gene coding for the muscle protein dystrophin; inherited as an X-linked recessive phenotype.

duplicate gene. An identical locus mapping to a second site in a genome.

duplication. More than one copy of a particular chromosomal segment in a chromosome set. Duplications supply genetic material capable of evolving new functions.

dyad. Two sister chromatids attached to the same centromere. A pair of sister chromatids joined at the centromere, as in mitosis and meiosis.

E

E (exit site). Site on the ribosome through which tRNAs pass after they have donated their amino acid to the growing nascent polypeptide chain.

E. coli. A common Gram-negative bacterium useful for cloning experiments. Present in human intestinal tract. Hundreds of strains of *E. coli* exist. One strain, K-12, has been completely sequenced.

ecdysone. A moulting hormone in insects.

ecogenetics. The branch of genetics that studies how (inherited or acquired) genetic factors influence human susceptibility to environmental health risks. It studies the genetic basis of environmental toxicity to develop methods for the detection, prevention and control of environment-related disease. Ecogenetics interacts with ecology, molecular genetics, toxicology, public health medicine and environmental epidemiology.

ecological genetics. The analysis of genetics of natural populations and of the adaptations of them to the environment.

ecology. The study of the interrelationships among living organisms and their environment. Human ecology means the study of human groups as influenced by environmental factors, including social and behavioral ones.

ectopic expression. The occurrence of gene expression in a tissue in which it is normally not expressed. Such ectopic expression can be caused by the juxtaposition of novel enhancer elements to a gene during genetic manipulation of transgenic organisms.

ectopic integration. In a transgenic organism, the insertion of an introduced gene (transgene) at a site other than its usual locus.

ediacaran fauna. E Vendian age assemblage of soft-bodied multicellular animals. The oldest fauna known.

effective population size (N or N_e). number of individuals contributing alleles to the next generation (N_f= number of mothers in a population; relevant in the calculation of number of generations for the fixation of a mitochondrial allele).

electrophoresis. (1) A technique for separating the components of a mixture of charged molecules (proteins, DNAs, or RNAs) in an electric field within a gel or other support. The movement of electrically charged molecules in an electric field often resulting in their separation. (2) See "Gel electrophoresis".

electroporation. A technique for transfecting cells by the application of a high-voltage electric pulse.

elongation factors (EF-Ts, EF-Tu, EF-G). Proteins necessary for the proper elongation and translocation processes during translation at the ribosome in prokaryotes. Replaced by eEF1 and eEF2 in eukaryotes.

embryo. A developing offspring during the period when most of its internal organs are forming. It is called fetus in the next stage of development.

embryonic (or tissue) polarity. The production of axes of asymmetry in a developing embryo or tissue primordium.

endogenote. Bacterial host chromosome. See merozygote.

endomitosis. Chromosomal replication without nuclear or cellular division that results in cells with many copies of the same chromosome.

endonuclease. (1) A nuclease which cuts a nucleic acid molecule by cleaving the phosphodiester bonds between two internal residues. Best known examples are restriction endonucleases. (2) An enzyme that cleaves or hydrolyzes phosphodiester bonds within a polynucleotide chain. (3) An enzyme which digests nucleic acids starting in the middle of the strand (as opposed to an exonuclease, which must start at an end). Examples include the restriction enzymes, DNase I and RNase A.

endopolyploidy. An increase in the number of chromosome sets caused by replication without cell division.

endosperm. Triploid tissue in a seed, formed from fusion of two haploid female and one haploid male nucleus.

enforced outbreeding. Deliberate avoidance of mating between relatives.

enhancer trap. A transgenic construction inserted in a chromosome which is used to identify tissue-specific enhancers in the genome. In such a construct, a promoter sensitive to enhancer regulation is fused to a reporter gene, such that expression patterns of the reporter gene identify the spatial regulation conferred by nearby enhancers.

enhancer. (1) A cis-acting (on either side of a gene) enhancer of promoter function. They are located 10 to 50 kb downstream or upstream of a gene. They may be tissue-specific. The enhancer effect is mediated through sequence-specific DNA-binding proteins (see also silencer). (2) A cis-regulatory sequence that can elevate levels of transcription from an adjacent promoter. Many tissue specific enhancers can determine spatial patterns of gene expression in higher eukaryotes. Enhancers can act on promoters over many tens of kilobases of DNA and can be 5' or 3' to the promoter they regulate. A eukaryotic DNA sequence that increases transcription of a region even if the enhancer is distant from the region being transcribed. (3) An enhancer is a nucleotide sequence to which transcription factor(s) bind, and which increases the transcription of a gene. It is NOT part of a promoter; the basic difference being that an enhancer can be moved around anywhere in the general vicinity of the gene (within several thousand nucleotides on either side or even within an intron), and it will still function. It can even be clipped out and spliced back in backwards, and will still operate. A promoter, on the other hand, is position- and orientation-dependent. Some enhancers are "conditional" - in other words, they enhance transcription only under certain conditions, for example in the presence of a hormone.

enriched medium. See complete medium.

enucleate cell. A cell having no nucleus.

environment. The combination of all the conditions external to the genome that potentially affect its expression and its structure.

environmental variance. The variance due to environmental variation.

enzyme. A protein that functions as a catalyst. Protein catalyst.

eocene epoch. A temporal subdivision (epoch) of the Tertiary period (58 to 37 Mya).

epidemiology. The study of the distribution and determinants of health-related events (including but not only disease epidemics) aiming to trace down their cause and subsequently to control health problems.

epigenesis. The theory that the development of an embryo consists of the gradual production and organization of parts.

epigenetics. The study of heritable changes in gene expression that occur without a change in DNA sequence. Epigenetic phenomena such as **imprinting** and **paramutation** violate Mendelian principles of heredity. Epigenetic studies link genotype to phenotype working out the chain of processes (mainly in **developmental biology**).

episome. A genetic element in bacteria that can replicate free in the cytoplasm or can be inserted into the main bacterial chromosome and replicate with the chromosome. Term used by Jacob and Wollman for genetic elements that can either exist independently in a cell or become integrated into the host chromosome.

epistasis. (1) The masking of the phenotypic effect of alleles at one gene by alleles of another gene. A situation in which the phenotypic expression of a genotype at one locus depends upon the genotype at another locus. (2) The nonreciprocal interaction between nonallelic genes. This may result in masking of one and in this case, the masked gene is said to be hypostatic. An epistatic-hypostatic relationship between two loci is similar to a dominant-recessive relationship between alleles at a particular locus.

equational division. The second meiotic division is an equational division because it does not reduce chromosome numbers. A nuclear division that maintains the same ploidy level of the cell.

equivalence group. A set of immature cells that all have the same developmental potential. In many cases cells of an equivalence group end up adopting different fates from one another.

ERE. Estrogen Response Element. A binding site in a promoter to which the activated estrogen receptor can bind. The estrogen receptor is essentially a transcription factor which is activated only in the presence of estrogens. The activated receptor will bind to an ERE, and transcription of the adjacent gene will be altered. See also "Response element".

escherichia coli. A gram-negative bacterium whose genome has been sequenced in its entirety. It is model organisms for the study of the prokaryotes.

ethidium. A molecule that can intercalate into DNA double helices when the helix is under torsional stress.

ethology. Study of animal behavior under normal conditions.

euchromatin. A chromosome region that stains poorly or not at all; thought to contain the normally functioning genes. Region of eukaryotic chromosome that is diffuse during interphase. Presumably the actively transcribing DNA of the chromosomes. (See heterochromatin)

eugenics. (1) Controlled human breeding based on notions of desirable and undesirable genotypes. (2) The idea of improving the quality of human species by selective breeding. Encouraging breeding of those with supposedly good genes is positive eugenics, whereas discouraging those with genes for undesirable traits is negative eugenics.

eukaryote. An organism having eukaryotic cells with true nuclei.

eukaryotic cell. (1) A cell containing a nucleus. (2) The DNA lies within a true nucleus (eu-karyon). May be unicellular (protist, some fungi) or multicellular (most fungi, plants, animals). Among eukaryotes, most fungi are haploid.

euploid. A cell having any number of complete chromosome sets, or an individual composed of such cells.

euploidy. The condition of a cell or organism that has one or more complete sets of chromosomes.

eutheria. Placental mammals. A subclass of the Class Mammalia (others are monotremes and marsupials). Embryo and fetus are nourished by a placenta. They are viviparous (producing young alive rather than lying eggs), have a long period of gestation and the young is born immature.

evolution. (1) In Darwinian terms a gradual change in phenotypic frequencies in a population that results in individuals with improved reproductive success. (2) The process that results in heritable changes in a population spread over many generations (change in allele frequencies over time). Biological evolution refers to populations and not to individuals and that the changes must be passed on to the next generations. Genes mutate, individuals are selected, and populations evolve.

evolutionarily stable strategy (ESS)**.** A strategy such that, if most members of a population adopt it, would give a reproductive fitness higher than any mutant strategy.

evolutionary rate. The rate of divergence between taxonomic groups measurable as amino acid substitutions per million years.

Ewens-Watterson Neutrality Test. Also called E-W homozygosity statistics. Described by Ewens (1972) and Watterson (1978). A widely used test in population genetics to estimate the selection acting on a locus. It compares the sum of observed homozygosity for each allele of a given locus (F_o) with the expected F_e value based on the number of alleles in the locus of interest, neutrality expectations and random mating assumption. A test of comparison yields an F value. Values close to zero mean that the locus is evolving under neutrality (genetic drift only) and there is no selection. Values of F significantly different from zero suggest selection. When $F_o > F_e$, the locus is undergoing purifying selection, and when $F_e > F_o$, the locus is under balancing selection (very common for HLA loci). Alternative tests for neutrality are Tajima's D (Tajima F: Genetics 1989; 123:585) and Slatkin's Exact Test for Neutrality (Slatkin M: Genet Res 1996; 68:259). See also Slatkin & Muirhead: Genetics 2000:156:2119).

excinuclease. The excision nuclease involved in nucleotide exchange repair of DNA.

excision repair. A process whereby cells remove part of a damaged DNA strand and replace it through DNA synthesis using the undamaged strand as a template. The repair of a DNA lesion by removal of the faulty DNA segment and its replacement with a new segment.

exconjugant. Each of the two cells that separates after conjugation has taken place. A female bacterial cell that has just been in conjugation with a male and that contains a fragment of male DNA.

exogenote. DNA that a bacterial cell has taken up through one of its sexual processes. See merozygote.

exon-intron boundary. Introns end with the dinucleotide ApG [3' splice site / acceptor] and start with the dinucleotide GpT [5' splice site / donor].

exon shuffling. The hypothesis put forward by Walter Gilbert that exons code for functional units of a protein and that evolution of new genes has proceeded by recombination or exclusion of exons.

exon. (1) A region of a gene that is present in the final functional transcript (mRNA) from that gene. Any non-intron section of the coding sequence of a gene; together the exons constitute the mRNA and are translated into protein. (2) The coding sequence of a eukaryotic gene (see also open reading frame). (3) Those portions of a genomic DNA sequence which WILL be represented in the final, mature mRNA. The term "exon" can also be used for the equivalent segments in the final RNA. Exons may include coding sequences, the 5' untranslated region or the 3' untranslated region.

exonuclease. (1) A nuclease which degrades a double-stranded DNA molecule by removing nucleotides from its two ends. (2) An enzyme that cleaves nucleotides one at a time from an end of a polynucleotide chain. An enzyme that hydrolyzes phosphodiester bonds from either the 3' or 5' terminus of a polynucleotide molecules. (3) An enzyme which digests nucleic acids starting at one end. An example is Exonuclease III, which digests only double-stranded DNA starting from the 3' end.

experimental design. A branch of statistics that attempts to outline the way in which experiments should be carried out so the data gathered will have statistical value.

exploitation competition. A form of competition that revolves around the superior ability to gather resources rather than an active interaction among organisms for these resources.

expressed sequence tag (EST). A unique DNA sequence derived from a cDNA library (therefore from a sequence which has been transcribed in some tissue or at some stage of development). The EST can be mapped, by a combination of genetic mapping procedures, to a unique locus in the genome and serves to identify that gene locus.

expression clone. This is a clone (plasmid in a bacteria, or maybe a l phage in bacteria) which is designed to produce a protein from the DNA insert. Mammalian genes do not function in bacteria, so to get bacterial expression from your mammalian cDNA, you would place its coding region (i.e. no introns) immediately adjacent to bacterial transcription/translation control sequences. That artificial construct (the "expression clone") will produce a pseudo-mammalian protein if put back into bacteria. Often, that protein can be recognized by antibodies raised against the authentic mammalian protein, and vice versa.

expression vector. A vector that has been designed to express cloned genes in a particular cell type.

expression. To "express" a gene is to cause it to function. A gene which encodes a protein will, when expressed, be transcribed and translated to produce that protein. A gene which encodes an RNA rather than a protein (for example, a rRNA gene) will produce that RNA when expressed.

expressivity. (1) The degree to which a particular genotype is expressed in the phenotype. The degree of expression of a genetically controlled trait. (2) The range of phenotypes resulting from a given genotype (cystic fibrosis, for example, may have a variable degree of severity). This is different from pleiotropy which refers to a variety of different phenotypes resulting from the same genotype.

extended phenotype. All effects of a gene upon the world where the effects influence the survival chance of a gene [Richard Dawkins].

extra-chromosomal inheritance. Non-Mendelian inheritance due to extra-nuclear DNA (mitochondrial DNA in animals). The transmission of the trait only occurs from mothers.

F

F' (F-prime) factor. Normally, the F factor contains genes related to conjugation/mating. The F' factor contains an additional portion of the bacterial genome.

F Factor (Fertility Factor). Transmissible plasmid (episome) in bacteria (such as E. coli) that acts as a sex factor. It is a circular DNA about 94 kb long. Conjugation and chromosomal gene transfer occur from F^+ (male) to F^- (female) bacterium.

F factor. See fertility factor.

F-strain. E.coli strain behaving as recipients during conjugation (female). It lacks the F factor.

F-(minus) cell. In *Escherichia coli*, a cell having no fertility factor; a female cell.

F'(prime) factor. A fertility factor into which a portion of the bacterial chromosome has been incorporated.

F+ strain. E.coli strain behaving as donors during conjugation (male). It has the F factor.

F+(plus) cell. In *Escherichia coli*, a cell having a free fertility factor; a male cell.

F1 generation. The first filial generation, produced by crossing two parental lines.

F_1. First filial (son or daughter) hybrids arising from a first cross. Subsequent generations are denoted by F_2, F_3 etc.

F2 generation. The second filial generation, produced by selfing or intercrossing the F1.

factorial. The product of all integers from the specified number down to one (unity).

facultative heterochromatin. Heterochromatin located in positions that are composed of euchromatin in other individuals of the same species, or even in the other homologue of a chromosome pair.

familial trait. A trait shared by members of a family.

family selection. A breeding technique of selecting a pair on the basis of the average performance of their progeny.

Fanconi's anemia. An heritable disease in human beings with a syndrome of congenital malformations (decreased blood cells, short stature, patchy dark skin pimentation, predisposition to cancer especially leukaemia) inherited as an autosomal recessive.

fate map. A map of an embryo showing areas that are destined to develop into specific adult tissues and organs. A map of the developmental fate of a zygote or early embryo showing the adult organs that will develop from material at a given position on the zygote or early embryo.

fauna. A certain species of animals occurring in a particular region or period.

f-duction. See sexduction.

fecundity selection. The forces acting to cause one genotype to be more fertile than another genotype.

feedback inhibition. A post-translational control mechanism in which the end product of a biochemical pathway inhibits the activity of the first enzyme of this pathway.

fertility factor (F factor). The plasmid that allows a prokaryote to conjugate with and pass DNA into an F- cell. A bacterial episome whose presence confers donor ability (maleness).

fertilization. Fusion of female and male haploid gametes to form a diploid zygote from which a new individual develops.

Filial generation. Offspring generation. F1 is the first offspring or filial generation; F2 is the second; and so on. Successive generations of progeny in a controlled series of crosses, starting with two specific parents (the P generation) and selfing or intercrossing the progeny of each new (F1; F2; . . .) generation.

filter enrichment. A technique for recovering auxotrophic mutants in filamentous fungi (in which non-auxotrophic organisms are filtered off leaving a residue of non-growing auxotrophs).

fimbriae. See pili.

fingerprint. 1. The characteristic spot pattern produced by electrophoresis of the polypeptide fragments obtained through denaturation of a particular protein with a proteolytic enzyme. 2. See DNA fingerprint.

fingerprinting. The use of RFLPs or repeat sequence DNA to establish a unique individual-specific pattern of DNA fragments.

first division segregation (FDS). The allele arrangement (4+4) of spores within an ordered ascus that indicates the lack of recombination between a locus and its centromere. A linear pattern (4+4) of spore phenotypes within an ordered ascus for a particular allele pair, produced when the alleles go into separate nuclei at the first meiotic division, showing that no crossover has occurred between that allele pair and the centromere. See Second division segregation.

Fisher's Fundamental Theorem. The rate of increase in fitness is equal to the additive genetic variance in fitness. This means that if there is a lot of variation in the population the value of S will be large.

Fisher's Theorem of the Sex Ratio. In a population where individuals mate at random, the rarity of either sex will automatically set up selection pressure favoring production of the rarer sex. Once the rare sex is favored, the sex ratio gradually moves back toward equality.

fitness (W). The relative reproductive success of a genotype as measured by survival; fecundity or other life history parameters. See Darwinian fitness and natural selection.

fitness. Lifetime reproductive success of an individual (i.e., the total number of offspring who themselves survive to reproduce). It can be seen as the extent to which an individual successfully passes on its genes to the next generation. It has two components: survival (viability) and reproductive success (fecundity). Variation in fitness is the major driving force in biological evolution.

five-prime (5') end. The end of a DNA or RNA strand with a free 5' phosphate group corresponding to the transcription initiation (see also three-prime end).

fixed allele. An allele for which all members of the population under study are homozygous, so that no other alleles for this locus exist in the population.

fixed breakage point. According to the heteroduplex DNA recombination model, the point from which unwinding of the DNA double helices begins, as a prelude to formation of heteroduplex DNA.

fixed. The establishment of a single allelic variant at a locus as a result random genetic drift.

flower. The part of an angiosperm containing the organs of reproduction (male stamen and female stigma as well as the ovary).

fluctuation test. A test used in microbial genetics to establish the random nature of mutation, or to measure mutation rates. An experiment by Luria and Delbruck that compared the variance in number of mutations among small cultures with subsamples of a large culture to determine the mechanism of inherited change in bacteria.

fMET. See formylmethionine.

focus map. A fate map of areas of the Drosophila blastoderm destined to become specific adult structures, based on the frequencies of specific kinds of mosaics.

Fokker-Planck equation. An equation that describes diffusion processes. It is used by population geneticists to describe random genetic drift.

footprinting, DNAase. DNA with protein bound is resistant to digestion by DNAase. When a sequencing reaction is performed using such DNA, a protected area representing the footprint of the bound protein will be detected. This permits identification of the protein binding regions of the DNA.

footprinting. (1) A technique by which one identifies a protein binding site on cellular DNA. The presence of a bound protein prevents DNase from "nicking" that region, which can be detected by an appropriately designed gel. (2) A technique to determine the length of nucleic acid in contact with a protein. While in contact the free DNA is digested. The remaining DNA is then isolated and characterized.

formylmethionine(fmet). A specialized amino acid that is the very first one incorporated into the polypeptide chain in the synthesis of proteins in Prokaryotes.

forward mutation. A mutation that converts a wild-type allele to a mutant allele. See also reversion.

founder effect (Sewall Wright effcct). (1) A type of genetic drift in which allele frequencies are altered in a small population which is a nonrandom sample of a larger (main) population. (2) Genetic drift observed in a population founded by a small non representative sample of a larger population.

f-pili. Sex pili. Hair-like projections on an F+, an F' or Hfr bacterium involved in anchorage during conjugation and presumably through which DNA passes.

fragile site. A chromosomal region that has a tendency to break.

fragile-X syndrome. The most common form of inherited mental retardation. Named for its association with an X chromosome with a tip that breaks or appears uncondensed. Inheritance involves imprinting.

frameshift mutation. (1) The insertion or deletion of a nucleotide pair or pairs, causing a disruption of the translational reading frame. (2) Mutations, usually deletions or insertions, that change the reading frame of the codon triplets.

frameshift. A mutation in which there is an addition or deletion of one, two or a small number (not a multiple of three) of nucleotides that causes the codon reading frame to shift to one of two others from the point of the mutation during translation. Consequently the amino acid sequence of the protein is altered from the point of the mutation to the carboxy terminus.

frequency histogram. A step curve in which the frequencies of various arbitrarily bounded classes are graphed.

frequency-dependent fitness. Fitness differences whose intensity changes with changes in the relative frequency of genotypes in the population.

frequency-dependent selection. Selection that involves frequency-dependent fitness. Selection of a genotype depending on its frequency in the population.

frequency-independent selection. Selection in which the fitnesses of genotypes are independent of their relative frequency in the population.

frequency-interdependent fitness. Fitness that is not dependent upon interactions with other individuals of the same species.

fruit. Mature ovary with seeds inside. Its function is seed protection and dispersal. Fruits are a development of the ovary wall and sometimes the other flower parts as well. Its formation is induced by the plant hormone auxin which is released by the maturing seeds.

fruiting body. In fungi, the organ in which meiosis occurs and sexual spores are produced.

Fu/HC. The fusion/histocompatibility system of the **Ascidians**. It is involved in self - nonself recognition regulating the fusion between compatible organisms and prevention of self-fertilization.

fugu. The puffer fish, Fugu rubripes, has essentially the same number of genes as the human genome, but its genome is eight times more compact than human genome (about 400 Mb as opposed to 3 Gb). A project to sequence the whole Fugu genome is underway.

functional alleles. Mutants that fail to complement each other in a cis-trans complementation test.

fundamental number. The number of chromosome arms in a somatic cell of a particular species.

fungus. A Kingdom made up of a diverse group of unicellular or multicellular, eukaryotic organisms which are not plants or animals. Many are parasitic or saprophytic. Both asexual and sexual reproductions are possible. The Kingdom includes five phyla: Zygomycetes (conjugating fungi, black bread molds), Deuteromycetes (reproduce only asexually, Aspergillus 'brown mold' and Penicillium), Basidiomycetes (incl. mushrooms), Ascomycetes (incl. Neurospora 'bread mold' and Saccharomyces 'baker's yeast') and Mycophycophyta (incl. lichens). Some of them (Basidiomycetes) have one of the most ancient pheromone-based mating-type recognition systems. See also dikaryosis, heterokaryon and mating types.

G

G. Guanine or guanosine.

Gaia hypothesis. A hypothesis developed by Lovelock, Watson, Margulis and others. In its form, -based on a mathematical model- this theory proposes that the physical and chemical condition of the surface of the Earth, of the atmosphere, and of the oceans has been and is actively made fit and comfortable by the presence of life itself. This is in contrast to the conventional wisdom which held that life adapted to the planetary conditions and they evolved their separate ways. Lovelock and Watson developed the Daisyworld model - an imaginary planet, which maintains conditions for its survival simply by following its own natural processes. This simple model has since become an integral part of the debate about the Gaia Hypothesis. The Daisyworld planet contains only two species of life: light daisies and dark daisies and an example of a self-regulating system.

gain-of-function dominant. A mutation in which dominance is caused by changing the specificity or expression pattern of a gene or gene product, rather than simply by reducing or eliminating the normal activity of that gene or gene product.

Galton's Regression Law. Individuals differing from the average character of the population produce offspring which, on the average, differ to a lesser degree but in the same direction from the average as their parents.

gamete. (1) A germ cell having a haploid chromosome complement. Gametes from parents of opposite sexes fuse to form zygotes. A specialized haploid cell that fuses with a gamete from the opposite sex or mating type to form a diploid zygote; in mammals, an egg or a sperm. (2) A haploid reproductive cell such as sperm (or pollen) and egg (oocyte).

gametic association. See linkage disequilibrium.

gametic selection. The forces acting to cause differential reproductive success of one allele over another in a heterozygote.

gametophyte. (1) The haploid stage of a plant life cycle that produces gametes (by mitosis). It alternates with a diploid sporophyte generation. The haploid gamete-producing stage in the life cycle of plants; prominent and independent in some species but reduced or parasitic in others. (2) The haploid, gamete-forming (sexual) generation in plants with alternation of generations. Typically it is produced from a haploid spore. See also sporophyte.

gastrulation. The process of movements and infoldings of embryonic cells destined to become endoderm in early animal embryos, immediately following blastula (or blastoderm) stage, generating the blastopore.

g-banding. Technique for producing banding patterns in eukaryotic chromosomes. Bands are produced by staining with Giemsa stain after pretreating chromosomes with trypsin. Each homologous chromosome pair has a unique pattern of g-bands, enabling recognition of particular chromosomes.

gel electrophoresis. A method to analyze the size of DNA (or RNA) fragments. In the presence of an electric field, larger fragments of DNA move through a gel slower than smaller ones. If a sample contains fragments at four different discrete sizes, those four size classes will, when subjected to electrophoresis, all migrate in groups, producing four migrating "bands". Usually, these are visualized by soaking the gel in a dye (ethidium bromide) which makes the DNA fluoresce under UV light.

gel shift assay. (Aka gel mobility shift assay (GMSA), band shift assay (BSA), electrophoretic mobility shift assay (EMSA)). A method by which one can determine whether a particular protein preparation contains factors which bind to a particular DNA fragment. When a radiolabeled DNA fragment is run on a gel, it shows a characteristic mobility. If it is first incubated with a cellular extract of proteins (or with purified protein), any protein-DNA complexes will migrate slower than the naked DNA - a shifted band.

gender. Differences between any two complementary organisms of the same species that render them capable of mating (see mating types).

gene amplification. A process by which the cell increases the number of a particular gene within the genome. The process by which the number of copies of a chromosomal segment is increased in a cell.

gene cloning. The production of a lineage of cells all of which contain one kind of DNA fragment of interest derived from a population of many kinds of DNA fragments. Operationally by:

inserting (recombining) a population of DNA molecules, known to contain the DNA of interest, into a population of vector DNA molecules in such a way that each vector molecule contains only a single DNA molecule from the original population;

transforming a population of host cells with the vector DNA recombinants such that each host cell takes up only one vector;

growing single host cells separately (cloning) by plating at low density to form a collection of separate colonies;

screening the colonies (clones) formed for the presence of the DNA of interest.

gene conversion. (1) A meiotic process of directed change in which one allele directs the conversion of a partner allele to its own form. In asci of Ascomycete fungi a 4:4 ratio of alleles is expected after meiosis, yet 6:2 and 5:3 ratios are sometimes observed. A model of recombination, produced by Holliday, suggests that gene conversion may be explained by repair of heteroduplex DNA. (2) Partial sequence transfer from one allele to another (interallelic recombination) converting one gene or allele to another one. It is the most common mechanism, especially for the HLA-B locus, in the generation of new MHC alleles. Less common are conversions between alleles of different MHC loci (intergenic conversion).

gene dose. The number of copies of a particular gene present in the genome.

gene family. A set of genes in one genome all descended from the same ancestral gene. A group of genes that has arisen by duplication of an ancestral gene. The genes in the family may or may not have diverged from each other.

gene flow. (1) The movement of genes from one population to another by way of interbreeding of individuals in the two populations. (2) The movement of genes within a population or between two populations following genetic admixture. Gene flow creates new combinations of genes or alleles in individuals that can be tested against the environment. This way it is one of the sources of variation in the process of natural selection.

gene frequency. See allele frequency.

gene fusion. The accidental joining of DNA of two genes, such as can occur in a translocation or inversion. Gene fusions can give rise to hybrid proteins or to the misregulation of the transcription of one gene by the cis regulatory elements (enhancers) of another.

gene interaction. The collaboration of several different genes in the production of one phenotypic character (or related group of characters).

gene locus. The specific place on a chromosome where a gene is located.

gene map. 1. A linear designation of mutant sites within a gene, based upon the various frequencies of interallelic (intragenic) recombination. 2. The DNA sequence of a gene annotated with sites of regulatory elements, introns, exons and mutations.

gene mutation. Mutation (point or larger change) that results from changes within the structure of a gene.

gene pair. The two copies of a particular gene present in a diploid cell (one in each chromosome set).

gene pool. All of the alleles available among the reproductive members of a population from which gametes can be drawn.

gene therapy. The correction of a genetic deficiency in a cell by the addition of new DNA and its insertion into the genome.

gene. (1) A unit of DNA which performs one function. Usually, this is equated with the production of one RNA or one protein. A gene contains coding regions, introns, untranslated regions and control regions. (2) Physical and functional unit of heredity that carries information from one generation to the next. The entire DNA sequence necessary for the synthesis of a functional polypeptide or RNA molecule. In addition to the coding regions (exons), a gene may have non-coding intervening sequences (introns) and transcription-control regions. (3) Segregating and heritable determinant of the phenotype. The fundamental physical and functional unit of heredity, which carries information from one generation to the next. A segment of DNA, composed of a transcribed region and regulatory sequences that make possible transcription. See also cistron; locus; Mendel's first law.

generalized transduction. The ability of certain phages to transduce any gene in the bacterial chromosome. Form of transduction in which any region of the host genome can be transduced. See specialized transduction.

genetic anticipation. The progressive shift of the age of onset of a hereditary disease to earlier ages in successive generations. It may occur because a parent is a mosaic, and the child has the full mutation in all cells. Triplet repeat expansion may demonstrate anticipation when the number of repeats increases with each generation.

genetic code. linear sequences of three nucleotides (triplets) that specify amino acids or termination (nonsense) codons during the process of translation at the ribosome. The correspondence between nucleotide triplets in DNA and amino acids in protein.

genetic dissection. The use of mutation and recombination to piece together the various components of a given biological function.

Genetic Distance Estimation by PHYLIP. The most popular (and free) phylogenetics program PHYLIP can be used to estimate genetic distance between populations. Most components of PHYLIP can be run online. One component of the package GENDST estimates genetic distance from allele frequencies using one of the three methods: Nei's, Cavalli-Sforza's or Reynold's (see papers by Nei et al, 1983, Nei M, 1996 and a lecture note for more information on these methods). GENDST can be run online using the default options (Nei's genetic distance) to obtain genetic distance matrix data. The PHYLIP program CONTML estimates phylogenies from gene frequency data by maximum likelihood under a model in which all divergence is due to genetic drift in the absence of new mutations (Cavalli-Sforza's method) and draws a tree. The program comes as a freeware as part of PHYLIP or this program can be run online with default options. If new mutations are contributing to allele frequency changes, Nei's method should be selected on GENDST to estimate genetic distances first. Then a tree can be obtained using one of the following components of PHYLIP: NEIGHBOR also draws a phylogenetic tree using the genetic distance matrix data (from GENDST). It uses either Nei's "Neighbor Joining Method," or the UPGMA (**u**nweighted **p**air **g**roup **m**ethod with **a**rithmetic mean; average linkage clustering) method. Neighbor Joining is a distance matrix method producing an unrooted tree without the assumption of a clock (UPGMA does assume a clock). NEIGHBOR can be run online. Other components of PHYLIP that draw phylogenetic trees from genetic distance matrix data are FITCH/online (does not assume evolutionary clock) and KITSCH/online (assumes evolutionary clock).

genetic distance. A measurement of genetic relatedness of populations. The estimate is based on the number of allelic substitutions per locus that have occurred during the separate evolution of two populations.

genetic drift. Evolutionary change over generations due to random events in small populations (not to be mixed with sampling error due to a small sample size). It operates unless overcome by strong selective forces. Wildly different HLA allele frequencies among South AmerIndian tribes are believed to be result of probable genetic drift in each small tribe.

genetic engineering. Popular term for recombinant DNA technology.

genetic fine structure. The structure of the gene analyzed at the level of the smallest units of recombination and mutation (nucleotides).

genetic heterogeneity. Presence of several different genotypes contributing to the genetic component of a disease on their own.

genetic linkage. The situation referring to segregation of two or more genes together as a unit. Genetic linkage is thought to arise to accommodate genes that function best in each other's company, i.e., to provide a necessary cooperative effect that enhances survival. Genetic linkage reflects a lack of meiotic crossovers between two genes.

genetic load. (1) The average number of lethal equivalents (or any recessive mutant lowering fitness) per individual in a population which are propagated by heterozygotes in a masked state. (2) The relative decrease in the mean fitness of a population due to the presence of genotypes that have less than the highest fitness.

genetic map. A linear designation of sites within a chromosome or genome, based upon the various frequencies of recombination between genetic markers. See linkage map

genetic marker. Allele, DNA marker or cytogenetic marker used as experimental probe to keep track of an individual, a tissue, a cell, a nucleus, a chromosome, or a gene.

genetic polymorphism. The occurrence together in the same population of more than one allele or genetic marker at the same locus with the least frequent allele or marker occurring more frequently than can be accounted for by mutation alone.

genetic variance. Phenotypic variance resulting from the presence of different genotypes in the population.

genetics. (1) The study of genes through their variation. (2) The study of inheritance. (3) Study of variation and heredity and their physical basis in DNA.

genic balance theory. The theory of Bridges that the sex of a fruit fly is determined by the relative number of X chromosomes and autosomal sets.

genocopy. The same phenotype due to different genetic causes.

genome. (1) The entire complement of genetic material in a chromosome set. The entire genetic complement of a prokaryote, virus, mitochondrion or chloroplast or the haploid nuclear genetic complement of a eukaryotic species. (2) The total DNA contained in each cell of an organism. Mammalian genomic DNA (including that of humans) contains 6x109 base pairs of DNA per diploid cell. There are somewhere in the order of a hundred thousand genes, including coding regions, 5' and 3' untranslated regions, introns, 5' and 3' flanking DNA. Also present in the genome are structural segments such as telomeric and centromeric DNAs and replication origins, and intergenic DNA. (3) Total genetic material in a set of haploid chromosomes as in a germ cell. The human genome contains 3,000 Mbp whereas the E.coli genome has 4.6 Mbp (see also C value).

genomic blot. A type of Southern blot specifically used to analyze a mixture of DNA fragments derived from total genomic DNA. Because genomic DNA is very complicated, when it has been digested with restriction enzymes, it produces a complex set of fragments ranging from tens of bp to tens of thousands of bp. However, any specific gene will be reproducibly found on only one or a few specific fragments. A million identical cells will produce a million identical restriction fragments for any given gene, so probing a genomic Southern with a gene-specific probe will produce a pattern of perhaps one or just a few bands.

genomic clone. A piece of DNA taken from the genome of a cell or animal, and spliced into a bacteriophage or other cloning vector. A genomic clone may contain coding regions, exons, introns, 5' flanking regions, 5' untranslated regions, 3' flanking regions, 3' untranslated regions, or it may contain none of these...it may only contain intergenic DNA (usually not a desired outcome of a cloning experiment!).

genomic imprinting. Differing expression of genetic material dependent on the parent-of-origin. This is due to methylation of one of the alleles depending of its origin. A very illustrative example is the inherited neck tumor paraganglioma for which the susceptibility gene is only active if inherited from the father. Genomic imprinting must be considered in disorders that appear to have skipped a generation.

genomic library. A set of cloned fragments representing the entire genome of an organism

genophore. The chromosome (genetic material) of prokaryotes and viruses.

genotype. (1) The diploid genetic formula at one or more loci. (2) The specific allelic composition of a cell, either of the entire cell or more commonly for a certain gene or a set of genes. The genes that an organism possesses.

geological timescale. The period between the origin of earth (4,500 Mya) and the beginning of the Cambrian period (540 Mya) is called the Precambrian Eon. The last 540 million years (Phanerozoic Eon) are divided into three eras: Palaeozoic (540-245 Mya); Mesozoic (245-65 Mya); Cainozoic. The geological periods (included in an era, longer than en epoch) are as follows: Vendian (immediately before the Cambrian; 610-540 Mya); Cambrian (540-510 Mya); Ordovician; Silurian; Devonian; Carboniferous; Permian; Triassic / Jurassic / Cretaceous (altogether the Mesozoic Era); Tertiary (65-1.64 Mya) and Quaternary. An epoch is a subdivision of a period.

germ line. Genetic material transmitted from one generation to the next through the gametes. A germ line mutation exists in all cells of the offspring formed from that gamete.

germinal mutation. mutation occurring in cells that are destined to develop into gametes.

germ-line theory. A theory to account for the high degree of antibody variability found in population. The germ-line theory suggests that every B lymphocyte has all the genes for every type of immunoglobulin but transcribes only one. See somatic mutation theory.

Giemsa stain. A complex of stains specific for the phosphate groups of DNA.

Gln. Glutamine (an amino acid).

Glu. Glutamate (an amino acid).

Gly. Glycine (an amino acid).

gnathastomata. Jawed vertebrates, evolved following the jawless vertebrates (Class Agnatha). The oldest extant branch of jawed vertebrates is the cartilaginous fish.

Goldstein-Hogness box. See TATA box.

gradient. A gradual change in some quantitative property over a specific distance.

gray crescent. A cortical region of the newly fertilised egg of frogs and some salamanders that forms just after fertilization on the side opposite sperm penetration.

GRE (Glucocorticoid Response Element). A binding site in a promoter to which the activated glucocorticoid receptor can bind. The glucocorticoid receptor is essentially a transcription factor which is activated only in the presence of glucocorticoids. The activated receptor will bind to a GRE, and transcription of the adjacent gene will be altered. See also "Response element".

Great Ape. Chimpanzees (including bonobos), gorillas, and orangutans.

ground state. The developmental state of a cell (or group of cells) in the absence of activation of a developmental regulatory switch.

group 1 intron. Self-splicing intron that requires an external guanine-containing nucleotide for splicing; releases the intron in a linear form.

group II intron. Self-splicing intron that does not require an external nucleotide for splicing; releases the intron in a lariat form.

group selection. Selection for traits that would be beneficial to a population at the expense of the individual possessing the trait.

GTP (guanosine 5'-triphosphate). A nucleotide that is a precursor in RNA synthesis which plays a role in protein synthesis (as well as in signal transduction and microtubule assembly). See also cap.

guanine. A purine base that pairs with cytosine. See purines.

guanosine. The nucleoside having guanine as its base.

guide RNA (gRNA). RNA that guides the insertion of uridines (RNA editing) into mRNAs in trypanosomes. Found in transcripts from minicircles and maxicircles of DNA in kinetoplasts.

gymnosperm (Gr. gymnos=naked; sperm=seed). Woody plants whose life histories include alternation of generations and ovules are not enclosed in a carpel. The pollen typically germinates on the surface of the ovule. A superclass in the sperm plants (Spermatophyta) division. Examples are cycad, conifer and gingko. Cycads are the most primitive ones evolved in the Devonian period about 400 Mya. There are about 700 extant species in Gymnosperms.

gynandromorph. An individual that is a mosaic of male and female structures. The underlying cause is frequently sex chromosome mosaicism, such that some cells are chromosomal females while others are chromosomal males. Mosaic individuals having simultaneous aspects of both the male and the female phenotype.

gyrase. One of the bacterial DNA topoisomerases that functions during DNA replication to reduce molecular tension caused by supercoiling (supertwisting). DNA gyrase produces, then seals, double-stranded breaks.

H

H-2 complex. The major histocompatibility complex (MHC) of the mouse. It is the first MHC discovered in 1937 by Peter Gorer.

haemoglobin (hb). The oxygen-transporting blood cell protein in most animals.

haemophilia. A disease in which the blood fails to clot. The most common form, primarily affecting males, is caused by a mutation in a gene coding for a clotting protein (factor VIII) inherited as an X-linked recessive phenotype.

hair pin loop. Binding of complementary sequences to each other to form a hair pin loop (also called stem loop). If happens in a PCR primer, it will not function.

half-chromatid conversion. A type of gene conversion that is inferred from the existence of non-identical sister spores in a fungal octad showing a non-Mendelian allele ratio.

Hamilton's Rule (theory of kin selection). In an altruistic act, if the donor sustains cost C, and the receiver gains a benefit B as a result of the altruism, then an allele that promotes an altruistic act in the donor will spread in the population if $B/C > 1/r$ or $rB-C>0$ (where r is the coefficient of relatedness).

Hamilton's Altruism Theory. If selection favored the evolution of altruistic acts between parents and offspring, then similar behavior might occur between other close relatives possessing the same altruistic genes which were identical by descent. In other words, individual may behave altruistically not only to their own immediate offspring but to others such as siblings, grandchildren and cousins (as happens in the bee society).

haplodiploidy. The sex-determining mechanism found in some insect groups among which males are haploid and females are diploid.

haploid number (n). The number of chromosomes in the gamete after meiosis. In humans, the haploid number is 23.

haploid. The state of having one copy of each chromosome per nucleus or cell. A cell having one chromosome set, or an organism composed of such cells.

haploidization. Production of a haploid from a diploid by progressive chromosome loss.

haploinsufficiency. Situation where one normal copy of a gene alone is not sufficient to maintain normal function. It is observed as a dominant mutation on one allele (or deletion of it) resulting in total loss-of - function in a diploid cell because of the insufficient amount of the wild-type protein encoded by the normal allele on the other haplotype (see also dominant negative).

haplotype. (1) A set of closely linked genetic markers present on one chromosome which tend to be inherited together (not easily separable by recombination). Some haplotypes may be in linkage disequilibrium. (2) The particular combination of alleles in a linked group encoded by genes in close vicinity on the same chromosome.

Hardy-Weinberg equilibrium (HWE). (1) In an infinitely large population, gene and genotype frequencies remain stable as long as there is no selection, mutation, or migration. **For a bi-allelic locus where** the gene frequencies are p and q: $p^2+2pq+q^2 = 1$. (2) The stable frequency distribution of genotypes, AA, Aa, and aa, in the proportions p^2, $2pq$, and q^2 respectively (where p and q are the frequencies of the alleles, A and a) that is a consequence of random mating in the absence of mutation, migration, natural selection, or random drift.

harlequin chromosome. Sister chromatids that stain differently, so that one appears dark and the other light (harlequin-like).

HAT medium. A selection medium for hybrid cell lines; contains hypoxanthine; aminopterin; thymidine. Only cell lines expressing both hypoxanthine phosphoribosyl transferase (HPRT+) and thymidine kinase (TK+) can survive in this medium. Aminopterin inhibits de novo synthesis of nucleosides, while HPRT and TK supply them from hypoxanthine and thymidine.

heat-shock protein. One of a number of proteins appearing in a cell after the cell has been subjected to elevated temperatures.

Heat Shock Response. Heat shock response is ubiquitous and highly conserved defense mechanism for protection of cells from harmful conditions such as heat shock, UV irradiation, toxic chemicals, infection, transformation and appearance of mutant and misfolded proteins. Heat Shock Proteins (HSPs) also function as accessory molecules in antigen presentation. HSP70 genes are within the MHC in most vertebrates. High levels of HSP70 prevent stress-induced apoptosis, and may have a transforming potential.

helicase. (1) A protein that unwinds DNA at replication forks. (2) An enzyme that unwinds the double DNA helix near the replication fork before DNA polymerase acts on it. Replication fork moves from 3' to 5' of the leading strand. Unwinding is also necessary for DNA repair. Mutations in the helicase genes on chromosome 2q and 19q are one group of causes of the DNA repair defect xeroderma pigmentosum (an autosomal recessive disease). See also primosome.

helix-loop-helix. A protein structural motif characteristic of certain DNA-binding proteins.

helix-turn-helix (HLH) protein. A protein in which a domain which forms two helices separated by a loop that acts as a sequence-specific DNA binding domain. HLH proteins are thought to act as transcription factors.

helix-turn-helix motif. Configuration, found in some DNA-binding proteins, consisting of a recognition helix and a stabilizing helix separated by a short loop.

hemizygous gene. A gene present in only one copy in a diploid organism for example, X-linked genes in a male mammal.

hemizygous. (1) As in any X-linked trait in males, absence of a homologous counterpart for an allele. It may also result from deletion. (2) The condition of loci on the X chromosome of the heterogametic sex of a diploid species. Or more generally when one part of the genome, in a normally diploid species, is present in only one copy.

heredity. The biological similarity of offspring and parents.

heritability in the broad sense. See broad heritability.

heritability in the narrow sense. The proportion of phenotypic variance that can be attributed to additive genetic variance.

heritability. (1) A measure of the degree to which the variance in the distribution of a phenotype is due to genetic causes. In the broad sense it is measured by the total genetic variance divided by the total phenotypic variance. In the narrow sense it is measured by the genetic variance due to additive genes divided by the total phenotypic variance. (2) The proportion of the total phenotypic variance that is attributable to genetic causes (h^2= genetic variance / total phenotypic variance).

hermaphrodite. An individual with both male and female genitalia. (1) A plant species in which male and female organs occur in the same flower of a single individual (compare monoecious plant). (2) An animal with both male and female sex organs.

hermaphroditism. Having both male and female sexual organs in one individual. Most invertebrates and plants are hermaphrodites. Union of the gametes of the same individual (self-fertilization) is the most extreme example of inbreeding.

heterochromatin. Densely staining condensed chromosomal regions, believed to be for the most part genetically inert. chromatin that remains tightly coiled (and darkly staining) throughout the cell cycle. See euchromatin

heteroduplex analysis. Duplex DNA formed by strands from different sources, referred to as a heteroduplex, will have loops and bubbles in regions where the two DNAs differ. Electron microscopic observation (analysis) of this DNA has been a useful tool in recombinant DNA work.

heteroduplex DNA model. A model that explains both crossing over and gene conversion by assuming the production of a short stretch of heteroduplex DNA (formed from both parental DNAs) in the vicinity of a chiasma.

heteroduplex DNA. See hybrid DNA.

heteroduplex mapping. The use of heteroduplex analysis to determine the location of various insertions; deletions; or heterogeneities between two DNA molecules.

heteroduplex. A DNA double helix formed by annealing single strands from different sources; if there is a sequence difference between the strands, the heteroduplex may show single strand loops or bubbles (unpaired regions).

heterogametic sex. (1) The sex which has the two different sex chromosomes (XY). Human and Drosophila males are the heterogametic sex, whereas, in birds, moths, some fish and amphibians, females are the heterogametic sex (ZW). (2) The sex with (usually) two heteromorphic (differently shaped) sex chromosomes (for example, X and Y). During meiosis it produces two kinds of gametes in regard to these sex chromosomes.

heterogeneous nuclear mRNA (hnRNA). The original RNA transcripts found in eukaryotic nuclei before post-transcriptional modifications. A diverse assortment of RNA types found in the nucleus, including mRNA precursors (pre-mRNA) and other types of RNA.

heterogeneous nuclear RNA (hnRNA). RNA products immediately synthesized from the DNA template in the nucleus (sometimes called DNA-like RNA or dRNA). This RNA species has a short half-life, is very heterogeneous and very large (molecular weight in excess of 10^7). hnRNA molecules are processed to generate the mRNA molecules (molecular weight generally less than $2x10^6$) before leaving the nucleus.

heterokaryon test. A test for cytoplasmic organelle mutations (e.g. mitochondrial or chloroplast mutations), based on new associations (recombination) of phenotypes in cells derived from genetically marked haploid heterokaryons. Since the heterokaryons are haploid and produce haploid spores by a process that does not involve diploid nuclei normal genetic recombination does not occur. Therefore evidence of recombination is evidence of reassociation by some other means such as reassociation of an organelle with nucleus.

heterokaryon. (1) A cell containing more than one genetically different nuclei. Naturally occurs in fungi as long as their fungal (heterokaryon) incompatibility types are identical (see also dikaryotic). (2) A cell that contains two or more nuclei from different origins. A cell composed of two different nuclear types in a common cytoplasm.

heteromorphic chromosomes. A chromosome pair with some homology but differing in size, shape, or staining properties. Homologous chromosome pair which are not morphologically identical (e.g., the sex chromosomes).

heteroplasmon. A cell containing a mixture of genetically different cytoplasms, generally different mitochondria or different chloroplasts.

heteroplasmy. The existence within an organism of genetic heterogeneity within the populations of mitochondria or chloroplasts.

heterothallic fungus. A fungus species in which two different mating types must unite to complete the sexual cycle.

heterothallic. (1) A botanical term used for organisms in which the two sexes reside in different individuals. (2) Organisms (fungi, algae, plants) which can only undergo sexual reproduction with another bearing a different mating/compatibility type (self-incompatible). See also homothallic.

heterotroph. Organism requiring an organic form of carbon as a carbon source.

heterozygosity. (1) A measure of the genetic variation in a population; with respect to one locus, stated as the frequency of heterozygotes for that locus. (2) Presence of two different alleles at a locus in a diploid organism (see homozygosity). It is the result of inheritance of different alleles from parents.

heterozygote advantage. A selection model in which heterozygotes have the highest fitness.

heterozygote. An individual having a heterozygous gene pair. A diploid or polyploid with different alleles at a particular locus.

heterozygous DNA. See hybrid DNA.

heterozygous gene pair. A gene pair having different alleles in the two chromosome sets of the diploid individual for example, Aa or, A1A2.

hexaploid. A cell having six chromosome sets, or an organism composed of such cells.

Hfr. (1) A male bacterial cell that has the F factor integrated into its chromosome is an Hfr (high frequency of recombination) cell. Crosses between Hfr cells and F- females produce far more recombinant progeny than do crosses between F+ males and F- females. (2) High frequency of recombination. A strain of bacteria that has incorporated an F factor into its chromosome and can then transfer the chromosome during conjugation. In *Escherichia coli*, a cell having its fertility factor integrated into the bacterial chromosome; a donor (male) cell.

Himalayan. A mammalian temperature-dependent coat phenotype, generally albino with pigment only at the cooler tips of the ears, feet, and tail.

His. Histidine (an amino acid).

histocompatibility antigen. Antigen that determines acceptance or rejection of a tissue graft by the immune system.

histocompatibility gene. A gene encoding an histocompatibility antigen.

histone. (1) A type of basic protein that forms the unit around which DNA is coiled in the nucleosomes of eukaryotic chromosomes. Arginine and lysine rich basic proteins making up a substantial portion of eukaryotic nucleoprotein. (2) Basic proteins that are involved in the packing of DNA. They bind to the phosphate groups of DNA. There are five major types of histone proteins.

HLA complex. The human major histocompatibility complex (MHC). An HLA haplotype has been totally sequenced in 1999.

hnRNA. (1) Heterogeneous nuclear RNA; refers collectively to the variety of RNAs found in the nucleus, including primary transcripts, partially processed RNAs and snRNA. The term hnRNA is often used just for the unprocessed primary transcripts, however. (2) See heterogeneous nuclear RNA.

Hogness box. See TATA box.

holandric gene. A gene carried on the Y chromosome and therefore transmitted from father to son.

holandric trait. Trait controlled by a locus found only on the Y chromosome. Involves father to son transmission.

holoenzyme. The complete enzyme including all subunits. Often used in reference to RNA and DNA polymerases.

homeo-box. (1) A consensus sequence of about 180 base pairs discovered in homeotic genes in Drosophila. Also found in other developmentally important genes from yeast to human beings. A family of quite similar 180base-pair DNA sequences that encode a polypeptide sequence called a homeo-domain, a sequence-specific DNA binding sequence. While the homeobox was first discovered in all homeotic genes, it is now known to encode a much more widespread DNA-binding motif. (2) Conserved protein sequence which forms a DNA-binding domain in a class of transcription factors.

homeo-domain. An approximately sixty-amino acid protein domain translated from the homeo-box. A highly conserved family of protein domain sequences 60 amino acids in length found within a large number of transcription factors that can form a helix-turn-helix structure and bind DNA in a sequence-specific manner.

homeologous chromosomes. Partially homologous chromosomes, usually indicating some original ancestral homology.

homeosis. The replacement of one body part by another. Homeosis can be caused by environmental factors leading to developmental anomalies, or by mutation.

homeotic mutations (genes). Mutations (defining genes) with a phenotype in which a given cell develops along a pathway normally followed by a different cell type. Mutations that can change the fate of an imaginal disk in insect development.

hominid. A member of the Hominidae family.

homogametic sex. The sex with homomorphic (similarly shaped) sex chromosomes (for example, XX), producing only one kind of gamete in regard to the sex chromosomes.

homologous chromosomes. (1) Chromosomes that occur in pairs one having come from the male parent and the other from the female parent. They pair participate in crossing-over during meiosis. Homologous chromosomes contain the same array of genes but may contain different alleles at those loci. (2) Members of a pair of essentially identical chromosomes that synapse during meiosis. Chromosomes that pair with each other at meiosis or chromosomes in different species that have retained most of the same genes during their evolution from a common ancestor.

homologous recombination. Breakage and reunion between homologous lengths of DNA mediated in Escherichia coli by products of the genes RecA and RecBCD, and functionally equivalent genes in other organisms.

homologue. A member of a pair of homologous chromosomes.

homology. A similarity due to inheritance from a common ancestor (see also analogy). An example is mammals' back legs.

homomorphic chromosomes. Morphologically identical members of a homologous pair of chromosomes.

homoplasmy. The existence within an organism of only one type of plastid; usually referring to genetic identity of mitochondria or chloroplasts.

homothallic fungus. A fungus species in which a single sexual spore can complete the entire sexual cycle (compare heterothallic fungus).

homothallic. (1) A botanical term used for groups whose individuals are not of different sexes. (2) Organisms (fungi, algae, plants) which can undergo sexual reproduction with a similar strain including the self (self-compatible) (see also heterothallic).

homozygosity. Presence of two identical alleles at a locus in a diploid organism (see heterozygosity). It is the result of inheritance of identical alleles from both parents.

homozygote. An individual having a homozygous gene pair. A diploid or a polyploid with identical alleles at a locus.

homozygous gene pair. A diploid gene pair having identical alleles in both copies for example, AA or, aa.

hormone response element (HRE). An HRE is a cis-regulatory DNA sequence for a hormone that acts by binding to a receptor that can act as a transcription factor, that is a binding site for the hormone-receptor complex.

hormone. Chemical secreted by one type of cell and acting on a second type of cell.

host range. 1. The spectrum of strains of bacterial species that a given strain of phage can infect. 2. The range of cells that can act as a host to a virus or bacteriophage.

host strain (bacterial). The bacterium used to harbor a plasmid. Typical host strains include HB101 (general purpose *E. coli* strain), DH5a (ditto), JM101 and JM109 (suitable for growing M13 phages), XL1-Blue (general-purpose, good for blue/white lacZ screening). Note that the host strain is available in a form with no plasmids (hence you can put one of your own into it), or it may have plasmids present (especially if you put them there). Hundreds, perhaps thousands, of host strains are available.

hot spot. A region of a genome exhibiting an increased frequency of some genetic phenomenon such as mutation, recombination and transposition.

house-keeping genes. Genes which are constitutively expressed in most cells because they provide basic functions.

Htf island. *Hpa Tiny Fragment* island which are unmethylated CpG-rich regions in the genome. Eighty percent of these occur at or near genes, particularly housekeeping genes. Many of the MHC genes discovered were not near Htf islands.

Huntington's disease. A late but variable age onset lethal human disease of nerve degeneration. Inherited as an autosomal dominant phenotype. Shows imprinting individuals inheriting HD from their father show significantly earlier onset than those inheriting a maternal allele. Shows genetic anticipation, in that the severity increase and age of onset decreases in affected individuals of later generations in a pedigree. The HD gene, encoding a protein called huntingtin, has been cloned.

H-Y antigen. The histocompatibility Y-antigen, a protein found on the cell surfaces of male mammals.

hybrid DNA. DNA whose two strands have different origins.

hybrid dysgenesis. A syndrome of effects including sterility, mutation, chromosome breakage, and male recombination in the hybrid progeny of crosses between certain laboratory and natural isolates of Drosophila.

hybrid plasmid. A plasmid that contains an inserted piece of foreign DNA.

hybrid vector/hybrid vehicle. An episome or plasmid containing an inserted piece of foreign DNA.

hybrid vigor (heterosis). Unusual growth, strength, and health of heterozygous offspring from two less vigorous homozygous parents.

hybrid zone. Geographical region in which previously isolated populations that have evolved differences come into contact and form hybrids.

hybrid. (1) A heterozygote. (2) A progeny individual from any cross involving parents of differing genotypes. Offspring of unlike parents. (3) A duplex polynucleotide formed by hybridization of two single stranded polynucleotides of different origin. (4) The offspring of two distinct species.

hybridization in situ. Finding the location of a gene or gene product by adding specific radioactive or chemically tagged probes for the gene and detecting the location of the radioactivity or chemical on the chromosome or in the cell after hybridization.

hybridization. (1) The reaction by which the pairing of complementary strands of nucleic acid occurs. DNA is usually double-stranded, and when the strands are separated they will re-hybridize under the appropriate conditions. Hybrids can form between DNA-DNA, DNA-RNA or RNA-RNA. They can form between a short strand and a long strand containing a region complementary to the short one. Imperfect hybrids can also form, but the more imperfect they are, the less stable they will be (and the less likely to form). To "anneal" two strands is the same as to "hybridize" them. (2) The specific reassociation of complementary strands of nucleic acids.

hybridize. (1) To form a hybrid by performing a cross. (2) To anneal nucleic acid strands from different sources.

hybridoma. A cell resulting from the fusion of a spleen cell and myeloma cell. These cells can be cloned and maintained indefinitely in cell culture and produce monoclonal antibodies.

hydrogen bond. A weak bond involving the sharing of an electron with a hydrogen atom; hydrogen bonds are important in the specificity of base pairing in nucleic acids and in the determination of protein shape.

hydroxyapatite. A form of calcium phosphate that binds double-stranded DNA.

hypermorph. A mutant with more than the normal amount of some gene product.

hyperploid. Aneuploid containing a small number of extra chromosomes.

hypervariable locus. Locus with many alleles; especially those whose variation is due to variable numbers of tandem repeats.

hypervariable region. The part of a variable region that actually determines the specificity of an immunoglobulin.

hypha (plural hyphae). A thread-like structure (composed of cells attached end to end) that forms the main tissue in many fungus species.

hypomorph. A mutant with less than the normal amount of some gene product.

hypoploid. Aneuploid with a small number of chromosomes missing.

hypostatic gene. A gene whose expression is masked by an epistatic gene.

hypothesis. An unproven but testable scientific proposition. A theory is a statement with some confirmation.

hypotrich. A protozoan of the Ciliate order which reproduces sexually or by asexual binary division. Sexual reproduction can be via conjugation (nuclear exchange) or gamete fusion. In the former, multiple mating types are involved. In gametic fusion there are only two types one of which is the only source of intracellular organelles (see also mating type).

I

ichthyosaur. An extinct group of marine fish-like or porpoise-like reptiles abundant in Mesozoic seas.

identity by descent. The state of two alleles when they are identical copies of the same ancestral allele (autozygous).

idiogram. A photograph or diagram of the chromosomes of a cell arranged in an orderly fashion. See karyotype.

idiomorph. This term is used to describe the fungal mating types which are extremely dissimilar from each other and do not show homology between strains of the opposite sex (as opposed to the allelic relationship in most polymorphic systems). Also used as ideomorph.

idiotypic variation. Variation unique to an individual immunoglobulin molecule produced by the variable region of immunoglobulin genes.

idling reaction. The production of guanosine tetraphosphate (3' ppGpp 5', magic spot) by the stringent factor when a ribosome encounters an uncharged tRNA in the A site.

Ig. See immunoglobulin.

Ile. Isoleucine (an amino acid).

imago. An adult insect.

immune system. The cells and tissues involved in recognizing and attacking foreign substances within the body of an animal.

immunity. The ability of an organism to resist infection.

immunoglobulin (Ig). A general term for the kind of globular blood proteins that constitute antibodies. A tetrameric protein composed of two identical light chains and two identical heavy chains. Specific proteins produced by derivatives of B lymphocytes that interact with and help protect an organism from specific antigens.

immunoglobulin gene. A gene encoding one of the protein chains (light or heavy) of an immunoglobulin. The germline genes are activated by a special recombination process that joins a number of different gene segments together. In the case of immunoglobulin light chains there are three segments (V variable, J joining and C constant). In the case of immunoglobulin heavy chains there are four segments (V variable, D diversity, J joining and C constant). There are many different V segments (light chains ~300, heavy chains ~500), 4 J segments in both light and heavy chain genes, 12 D segments in the heavy chain gene and 1 C segment in light chain genes, 5 C segments in heavy chain gene. The recombination process appears to be able to join any V with any J (light chain) or any V to D to J (heavy chain). The total number of possible immunoglobulins (i.e. the number of possible light chains times the number of possible heavy chains) that can be generated by this mechanism is thus several 10s of millions (300x4x500x12x4=28800000). It is certain that the number is greater than this because the recombination events themselves are not precise and random loss of nucleotides from each recombining end occurs.

immunohistochemistry. The use of antibodies or antisera as histological tools for identifying patterns of antigen distribution within a tissue or an organism. An antibody (or mixture of antibodies) that binds to a specific protein or other antigen is tagged with a fluorescent chemical or an enzyme that can convert a substrate to a visible dye. The tagged antibody is incubated with the tissue and after washing unbound antibody away the bound antibody distribution is revealed by fluorescence microscopy or incubation with a chromogenic substrate. It is assumed that antibody distribution reflects antigen distribution.

imprinting. (1) See genomic imprinting. (2) See molecular imprinting.

***in situ*.** In place

***in vitro* mutagenesis.** The production of either random or specific mutations in a piece of cloned DNA. Typically, the DNA will then be reintroduced into a cell or an organism to assess the results of the mutagenesis.

***in vitro*.** (1) In an experimental situation outside the organism. Biological or chemical work done in the test tube (literally in glass) rather than in living systems. (2) Literally, 'in glass' meaning in the laboratory.

***in vivo*.** In a living cell or organism.

inbreeding coefficient (F). The probability of homozygosity by descent (having common ancestors). The probability that a zygote obtains copies of the same ancestral gene from both its parents because they are related. The probability of autozygosity.

inbreeding depression. A depression of vigour or yield due to inbreeding.

inbreeding depression. Reduction in offspring fitness resulting from mating between blood relatives.

inbreeding. (1) Production of offspring by (blood) related parents. Its most extreme form is self-fertilization in hermaphrodites (most invertebrates and plants). (2) The mating of genetically related individuals. Mating between relatives.

incest. Sexual relationships between parents and children, or between brothers and sisters.

incestuous. A mating between blood relatives who are more closely related than the law of the land allows.

inclusive fitness. The expansion of the concept of the fitness of a genotype to include benefits accrued to relatives of an individual since relatives share parts of their genomes. Hence an apparently altruistic act toward a relative may in fact enhance the fitness of the individual performing the act.

incomplete dominance. (1) One allele is not expressed, but the other allele expresses itself normally so that the phenotype gets half the dose of the effect. (2) The situation in which both alleles of a heterozygote influence the phenotype. The phenotype is usually intermediate between the two homozygous phenotypes. The situation in which a heterozygote shows a phenotype somewhere (but not exactly half-way) intermediate between the corresponding homozygote phenotypes. (Exact intermediacy is no dominance.) See also dominance, codominance and recessivity.

independent assortment. Mendel's second rule describing the independent segregation of alleles of different unlinked or remotely linked loci.

inducer. An environmental agent that triggers transcription from an operon.

inducible operon. A gene system, often encoding a coordinated group of enzymes involved in a catabolic pathway, is inducible if an early metabolite in the pathway causes activation, usually by interaction with and inactivation of a repressor, of transcription of the genes encoding the enzymes.

induction. 1. The relief of repression for a gene or set of genes under negative control by a repressor. 2. Regarding temperate phages, the process causing a prophage to become virulent. 3. In development, an interaction between two cell lineages to alter the developmental fate of one or both of them.

inductive interaction. The interaction between two groups of cells in which a signal passed from one group of cells causes the other group of cells to change their developmental state (or fate).

industrial melanism. The darkening of several species of moths during the recent period of industrialization in many countries by the spread of a gene controlling elevated melanin synthesis.

infectious transfer. The rapid transmission of free episomes (plus any host chromosomal genes they may carry) from donor to recipient cells in a bacterial population.

initiation codon. The mRNA sequence AUG, which specifies methionine, the first amino acid used in the translation process. (Occasionally GUG, valine, is recognized as an initiation codon).

initiation complex. (1) A multi-protein complex that forms at the site of transcription initiation and is composed of RNA polymerase II, ubiquitous or general transcription or initiation factors (TFII or IF/eIF) and gene-specific enhancers/silencers. (2) The complex formed for initiation of translation. It consists of the 30S ribosomal subunit; mRNA; N-formyl-methionine tRNA; and three initiation factors .

initiation factors (IF1, IF2, IF3). Proteins (prokaryotic with eukaryotic analogues) required for the proper initiation of translation.

initiator protein. Protein that recognizes the origin of replication on a replicon and takes part in primosome construction.

innate immunity. Pre-existing and non-specific defense immunity with a very low memory component if any. As the primitive immune response against bacteria, it is present in invertebrates and vertebrates.

inosine. A rare base that is important at the wobble (3rd) position of some tRNA anticodons.

insert. In a complete plasmid clone, there are two types of DNA - the "vector" sequences and the "insert". The vector sequences are those regions necessary for propagation, antibiotic resistance, and all those mundane functions necessary for useful cloning. In contrast, however, the insert is the piece of DNA in which you are really interested.

insertion mutagenesis. Change in gene activity due to an insertion (of DNA) event that either changes the gene directly or disrupts control mechanisms.

insertion sequence (IS). A mobile piece of bacterial DNA (several hundred nucleotide pairs in length) that is capable of inactivating a gene into which it inserts. Small simple transposons. See transposable genetic element.

insertional translocation. The insertion of a segment from one chromosome into another non-homologous chromosome.

inside marker. The middle locus of three linked loci.

integrase. An enzyme that catalyzes a site-specific recombination (integration or excision) involving a prophage and a bacterial chromosome.

intercalary heterochromatin. Heterochromatin, other than centromeric heterochromatin, dispersed throughout eukaryotic chromosomes.

intercalating agent. A chemical that can insert itself between the stacked bases at the centre of the DNA double helix, possibly causing a frameshift mutation.

interchromosomal recombination. Recombination resulting from independent assortment.

interference competition. A form of competition that involves a fight or other active interaction among organisms.

interference. A measure of the independence of crossover events from each other, calculated by subtracting the coefficient of coincidence from 1.

intergenic suppression. A mutation at a second locus that apparently restores the wild-type phenotype to a mutation at a first locus.

intergenic. Between two genes; e.g. intergenic DNA is the DNA found between two genes. The term is often used to mean non-functional DNA (or at least DNA with no known importance to the two genes flanking it). Alternatively, one might speak of the "intergenic distance" between two genes as the number of base pairs from the polyA site of the first gene to the cap site of the second. This usage might therefore include the promoter region of the second gene.

interkinesis. The abbreviated interphase that occurs between meiosis I and II. No DNA replication occurs here.

interphase. The cell cycle stage between nuclear divisions, when chromosomes are extended and functionally active. The metabolically active non-dividing stage of the cell cycle.

interrupted mating. A technique used to map bacterial genes by determining the sequence in which donor genes enter recipient cells. A gene mapping technique in which bacterial conjugation is disrupted after specified time intervals.

intersex. An organism with external sexual characteristics that have attributes of both sexes.

interstitial region. The chromosomal region between the centromere and the site of a rearrangement.

intra-allelic complementation. The restoration of activity or partial activity (complementation) to an enzyme made of identical subunits (polypeptides encoded by one gene) in a heterozygote of two different mutant alleles of that gene.

intrachromosomal recombination. Recombination resulting from crossing over between two linked gene pairs.

intragenic suppression. A second mutation **within** a mutant gene that results in an apparent restoration of the original phenotype.

intron (intervening sequence). A DNA segment of largely unknown function within a gene that specifically interupts the coding (exon) sequences of that gene. Introns are transcribed as part of the normal gene primary transcript, but intron sequences are not found in the functional mRNA. Intron sequences are removed from the primary transcript by a splicing mechanism.

intron. (1) A non-coding section of DNA within a gene that is not translated to a peptide. Intervening sequences between exons. Introns are featured in the primary transcript (pre-mRNA) but removed by splicing during nuclear RNA processing/editing. (2) Introns are portions of genomic DNA which **ARE**

transcribed (and thus present in the primary transcript) but which are later spliced out. They thus are not present in the mature mRNA. Note that although the 3' flanking region is often transcribed, it is removed by endonucleolytic cleavage and not by splicing. It is not an intron.

inversion. A chromosomal mutation involving the removal of a chromosome segment, its rotation through 180 degrees, and its reinsertion in the same location. The replacement of a section of a chromosome in the reverse orientation.

invertebrate. All animals other than those in the phylum Chordata; lower metazoans. They do not possess a notochord or vertebral column. Examples are worms, corals, sponges, etc. The protochordates are sometimes called higher invertebrates.

inverted repeat (IR) sequence. A sequence found in identical (but inverted) form, for example, at the opposite ends of a transposon. A nucleotide sequence found at two sites on the same double helix but in opposite orientation.

iojap. A locus in maize that produces variegation.

ionizing radiation. Radiation, such as X-rays and gamma rays (high energy photons), that causes atoms to release electrons and become ions.

IR. See inverted repeat sequence.

IS element. See insertion sequence

IS. See insertion sequence.

isoaccepting tRNA. The various types of tRNA molecule carrying a specific amino acid.

isochromosome. A chromosome with two genetically and morphologically identical arms.

isoschizomer. Restriction endonuclease that recognizes the same target DNA sequence and cleaves it in the same way.

isotope. One of several forms of an element having the same atomic number but differing atomic masses.

isozyme. A different electrophoretic form of the same multi-subunit enzyme. Unlike allozymes isozymes are due to differing subunit configurations rather than allelic differences.

iterative evolution. Repeated origination of lineages with generally similar morphology at different times in the history of a clade.

J

Jacobson's organ. In some vertebrates, an accessory olfactory organ developed in connection with the roof of the mouth.

jawless vertebrates (Class Agnatha). The most primitive Vertebrates evolved about 500 Mya. The extant species hagfish and lampreys in the order Cyclostomata belong to this group.

junctional diversity. Variability in immunoglobulins caused by differences in the exact crossover point during V-J, V-D, and D-J joining.

kappa particle. Bacteria-like particle that gives a *Paramecium* the killer phenotype.

karyokinesis. The process of nuclear division. See cytokinesis.

karyotype. The chromosome complement of a cell. The entire chromosome complement of an individual or cell, as seen during mitotic metaphase. See idiogram.

KB. Abbreviation for kilobase, one thousand bases.

kin recognition. Ability to distinguish the incest from unrelated members of a species.

kin selection. (1) Natural selection for behaviors that lowers an individual's own chance of survival but raises that of a relative. (2) The mode of natural selection that acts on an individual's inclusive fitness.

kinase. A kinase is in general an enzyme that catalyzes the transfer of a phosphate group from ATP to something else. In molecular biology, it has acquired the more specific verbal usage for the transfer onto DNA of a radiolabeled phosphate group. This would be done in order to use the resultant "hot" DNA as a probe.

kinetochore. The chromosomal attachment point for the spindle fibers located within the centromeres.

kingdom. The major taxonomic group in the current classification of living organisms with the exception of informal division of prokaryotic and eukaryotic empires. The five Kingdoms are Monera, Protoctista, Fungi, Plants and Animals. In the late 1980s Cavalier-Smith proposed that within the Eukaryota there are six kingdoms: Archezoa, Protozoa, Chromista, Plants, Fungi, and Animals (see also taxonomy).

Klinefelter syndrome. An abnormal human male phenotype involving an extra X chromosome (XXY).

knock-out experiment. A technique for deleting, mutating or otherwise inactivating a gene in a mouse. This laborious method involves transfecting a crippled gene into cultured embryonic stem cells, searching through the thousands of resulting clones for one in which the crippled gene exactly replaced the normal one (by homologous recombination), and inserting that cell back into a mouse blastocyst. The resulting mouse will be chimaeric but, if you are lucky (and if you've gotten this far, you obviously are), its germ cells will carry the deleted gene. A few rounds of careful breeding can then produce progeny in which both copies of the gene are inactivated.

Kozak sequence. In some viral mRNAs, the consensus sequence surrounding the initiating AUG 5' ACCAUGG 3'. It facilitates ribosomal binding and therefore, protein synthesis. The most consistent position is located three nucleotides before the initiation codon (ATG) and is almost always an adenine nucleotide (see also Shine-Dalgarno sequence).

L

***lac* operon.** (1) A structural unit in the E.coli genome that consists of three structural genes (encoding different enzymes involved in sugar metabolism) transcribed together and their common promoter and operator genes. Provides a good model for studying the interactions between promoters and repressors. (2) An inducible operon including three loci involved in the uptake and breakdown of lactose in Escherichia coli.

lagging strand. In DNA replication, the strand that is synthesized apparently in the 3' to 5' direction, but actually in the 5' to 3' direction by ligating short fragments synthesized individually. Strand of DNA being replicated discontinuously. See also leading strand.

lambda dgal. A lambda phage carrying a gal(galactose) bacterial gene and defective (d) for some phage function.

lambda phage. One kind (species) of temperate bacteriophage.

lambda. See Bacteriophage Lambda.

lampbrush chromosomes. Chromosomes of amphibian oocytes having loops suggestive of a lampbrush. Large chromosomes found in amphibian eggs, with lateral

DNA loops, extending from chromomeres, producing a brushlike appearance under the microscope.

last male sperm precedence. A situation that results in fertilization of the ovum by the sperm of the last male in multiply inseminated female. This is due to sperm incapacitation by the semen and sperm displacement. This well-documented form of sperm competition is best known in Drosophila.

lateral inhibition. The signal produced by one cell that prevents adjacent cells from acquiring their fate.

lawn. A continuous layer of bacteria on the surface of an agar medium.

leader peptide gene. A small gene within the attenuator control region of repressible amino acid operons. Translation of a specific leader peptide tests for the concentration of a specific amino acid or set of amino acids in the cell. If the concentration is high transcription is terminated in the attenuator, if low continued leading to the production of enzymes for amino acid synthesis and hence more amino acid. See leader transcript.

leader sequence. (1) A sequence at the 5' (N-terminal) of the DNA and mRNA which leads the newly synthesized mRNA to the ribosome (it is not translated). It is also used to mean the signal sequence which is translated but is subject to post-translation cleavage when the final destination is reached or following secretion. (2) The sequence at the 5' end of an mRNA that is not translated into protein. The length of untranslated mRNA from the 5' end to the initiation codon AUG.

leader transcript. The mRNA transcribed by the attenuator region of repressible amino acid operons. The transcript is capable of several alternative stem-loop structures dependent on translation of a short leader peptide gene.

leading strand. Strand of DNA being replicated continuously. In DNA replication, the strand that is made in the 5' to 3' direction by continuous polymerization at the 3' growing tip. See also lagging strand.

leaky mutant. A mutant (typically an auxotroph) that results from a partial rather than a complete inactivation of the wild-type function.

lek. A special site where communal courtship display takes place by swarms of animals. Lekking is best known in male birds to attract female mates. Rare form of lekking by females is seen in the African butterfly (Acraea encedon), the Dance fly (Empis borealis) and the European dotterel bird (Eudromias morinellus).

leptonema (leptotene stage). The first stage of prophase 1 of meiosis in which chromosomes become distinct.

lesion. A damaged area (a mutant site) in a gene, a chromosome, or a protein.

lethal gene. A gene whose expression results in the death of the individual expressing it.

lethal-equivalent alleles. alleles whose summed effect is that of lethality for example, four alleles each of which would be lethal 25% of the time (or to 25% of their bearers), are equivalent to one lethal allele.

Leu. Leucine (an amino acid).

leucine zipper. A motif found in certain proteins in which Leu residues are evenly spaced through an a-helical region, such that they would end up on the same face of the helix. Dimers can form between two such proteins. The Leu zipper is important in the function of transcription factors such as Fos and Jun and related proteins.

leucine zipper. Configuration of a DNA-binding protein in which leucine residues on two protein alpha-helices interdigitate in zipper fashion to stabilize the protein.

leukemia. Cancer of the bone marrow resulting in excess production of leukocytes.

level of significance. The probability value in statistics used to reject the null hypothesis.

library. A library might be either a genomic library, or a cDNA library. In either case, the library is just a tube carrying a mixture of thousands of different clones - bacteria or l phages. Each clone carries an "insert" - the cloned DNA.

A cDNA library is usually just a mixture of bacteria, where each bacteria carries a different plasmid. Inserted into the plasmids (one per plasmid) are thousands of different pieces of cDNA (each typ. 500-5000 bp) copied from some source of mRNA, for example, total liver mRNA. The basic idea is that if you have a large enough number of different liver-derived cDNAs carried in those bacteria, there is a 99% probability that a cDNA copy of any given liver mRNA exists somewhere in the tube. The real trick is to find the one you want out of that mess - a process called screening (see "Screening").

A genomic library is similar in concept to a cDNA library, but differs in three major ways - 1) the library carries pieces of genomic DNA (and so contains introns and flanking regions, as well as coding and untranslated); 2) you need bacteriophage l or cosmids, rather than plasmids, because... 3) the inserts are usually 5-15 kb long (in a l library) or 20-40 kb (in a cosmid library). Therefore, a genomic library is most commonly a tube containing a mixture of l phages.

Enough different phages must be present in the library so that any given piece of DNA from the source genome has a 99% probability of being present.

ligand-receptor interaction. interaction between a molecule (usually of an extracellular origin) and a protein on or within a target cell. One type of ligand-receptor interaction can be between steroid hormones and their cytoplasmic or nuclear receptors. Another can be between secreted polypeptide ligands and transmembrane receptors.

ligase. (1) An enzyme that can rejoin a broken phosphodiester bond in a nucleic acid. (2) An enzyme which is of vital importance in recombinant DNA technology. It joins nucleotides together by a phosphodiester bond between the 5'-*P* end of a polynucleotide chain and the 3'-*OH* end of another one. (3) An enzyme, T4 DNA ligase, which can link pieces of DNA together. The pieces must have compatible ends (both of them blunt, or else mutually compatible sticky ends), and the ligation reaction requires ATP.

ligation. The process of splicing two pieces of DNA together. In practice, a pool of DNA fragments are treated with ligase (see "Ligase") in the presence of ATP, and all possible splicing products are produced, including circularized forms and end-to-end ligation of 2, 3 or more pieces. Usually, only some of these products are useful, and the investigator must have some way of selecting the desirable ones.

line. A group of identical pure-breeding diploid or polyploid organisms, distinguished from other individuals of the same species by some unique phenotype and genotype.

LINE. Long interspersed element; a type of large repetitive DNA segment found throughout the genome of Eukaryotes.

linear tetrad. A tetrad that results from the occurrence of the meiotic and postmeiotic nuclear divisions in such a way that sister products remain adjacent to one another in an ordered manner constrained within an ascus.

LINES (long interspersed elements). One of the abundant intermediate (6 to 7 kb) repeat DNA sequences in mammals (see also SINES).

linkage disequilibrium (LD). The tendency for two 'alleles' to be present on the same chromosome (positive LD), or not to segregate together (negative LD). As a result, specific alleles at two different loci are found together more or less than expected by chance. The same situation may exist for more than two alleles. Its magnitude is expressed as the delta (?) value and corresponds to the difference between the expected and the observed haplotype frequency. It can have positive or negative values. LD is decreased by recombination. Thus, it decreases every generation of random mating unless some process opposing the approach to linkage equilibrium. Permanent LD may result from natural selection if some gametic combinations result in higher fitness than other combinations.

linkage disequilibrium. (1) When the observed frequencies of haplotypes in a population does not agree with haplotype frequencies predicted by multiplying together the frequency of individual genetic markers in each haplotype. (2) When the observed frequencies of haplotypes in a population agrees with haplotype frequencies predicted by multiplying together the frequency of individual genetic markers in each haplotype.

linkage group. A group of gene loci known to be linked; a chromosome. There are as many linkage groups as there are homologous pairs of chromosomes. See synteny.

linkage map. An abstract map of chromosomal loci, based on recombinant frequencies.

linkage. (1) An association of gene loci on the same chromosome. (2) The tendency of 'genes' on the same chromosome to segregate together. This means that linked genes are transmitted to the same gamete more than 50% of the time. Genetic linkage reflects a lack of meiotic crossovers between two genes.

linker. A small piece of synthetic double-stranded DNA which contains something useful, such as a restriction site. A linker might be ligated onto the end of another piece of DNA to provide a desired restriction site.

linking number. The number of times one strand of a helix coils about the other.

liposome. Small spheres of phospholipid bilayers (just like cell membranes). They are used for gene transfer into cells. The DNA to be transferred is attached to them. The DNA-liposome complex fuses with the cell membrane to enter the cell and releases the DNA into the cell.

living fossil. An extant species which is morphologically very similar to a species from the ancient past. Despite apparent lack of change, they seem to have escaped extinction. Coelacanth (a 350 million-years-old lobe-finned fish), Horseshoe Crab (a 510 million-years-old marine arthropod), Amazon River Dolphin, Gingko (maidenhair tree, a gymnosperm), and Metasequoia (Metasequoia glyptostrobodes, a conifer) are examples.

lobe-finned fish. A group of fish that have bone and muscle in their limbs as opposed to simple fins as in most teleosts. The group includes coelacanth and lungfish.

locus(Plural loci). (1) The position of a gene, DNA marker or genetic marker on a chromosome. See gene locus. (2) The position on a chromosome occupied by a particular gene (plural: loci).

lod score method. A technique (logarithmic odds) for determining the most likely recombination frequency between two loci from pedigree data.

lod score. Logarithm of the odds favoring linkage obtained from the statistical analysis of linkage. The lod score (Z) of +3 is considered evidence for linkage.

lymphoma. cancer of the lymph nodes and spleen that causes excessive production of lymphocytes.

Lyon hypothesis. The hypothesis, named after Mary Lyon who stated it, suggesting that doseage compensation in mammals is by inactivation of all but one X chromosome in cells with more than one X chromosome. The Barr body, visible in some female mammalian cells, is an inactivated X chromosome.

Lyon hypothesis. The proposition by Mary F Lyon that random inactivation of one X chromosome in the somatic cells of mammalian females is responsible for **dosage compensation** and mosaicism.

Lys. Lysine (an amino acid).

lysate. The contents released from a lysed cell.

lysis. The breaking open of a cell by the destruction of its wall or membrane. The rupture and death of a bacterial cell upon the release of phage progeny.

lysogen. See lysogenic bacterium.

lysogenic bacterium. A bacterial cell capable of spontaneous lysis due, for example, to the release of a prophage from the bacterial chromosome.

lysogenic. The state of a bacterial cell that has an integrated phage (prophage) in its chromosome.

M

m.u. See map unit.

M13. A bacteriophage which infects certain strains of *E. coli* . The salient feature of this phage is that it packages only a single strand of DNA into its capsid. If the investigator has inserted some heterologous DNA into the M13 genome, copious quantities of single-stranded DNA can subsequently be isolated from the phage capsids. M13 is often used to generate templates for DNA sequencing.

macromolecule. A large polymer such as DNA, RNA, protein, lipid or polysaccharide.

maize (Zea mays). A crop plant used as animal and human food. It features frequently in scientific literature because it is the only diploid plant among economically important food crops such as oats and wheat (which are polyploids). Its disadvantages are that it has a relatively long generation time (a maximum of three generations per year), and a very large and complex genome with a lot of repetitive sequences. Natural cross-pollination ability of maize is another problem. Its meiotic chromosomes are, however, excellent subjects for cytogenetic studies. At present, the model system in plant genetics is Arabidopsis thaliana.

Major Histocompatibility Complex (MHC). (1) A genetic complex of vertebrates consisting of around 100 genes including the extremely polymorphic cell surface molecules called HLA in humans and H-2 in mice. These molecules provide an immunological marker for selfness and a genetic self-identity to the individual. This information is used in mate choice, union of gametes, maintenance of pregnancy, and immune response against nonself (including a transplanted graft). These molecules are the most polymorphic ones in vertebrates. The polymorphism arises from point mutations not at an unusually higher rate than other genes, and mainly from interallelic gene conversion events. The polymorphism is maintained through pathogen and non-pathogen driven mechanisms via heterozygous advantage (overdominant selection) and negative frequency dependent selection. A 3.6 Mb long human MHC haplotype and the 92 kb chicken MHC have been totally sequenced (see Nature 1999 (*Oct 28*); 401:921-925). (2) A group of highly polymorphic genes whose products appear on the surface of cells imparting the property of self (belonging to that organism). A genetic region found in all mammals whose products are primarily responsible for the rapid rejection of tissue grafts between indiviuals, and function in signalling between lymphocytes and cells expressing antigen.

mammals. One of the eight Classes in the Phylum Chordata which contains approximately 4500 species in 15 Orders. In mammals, the fertilization of the egg is internal, the young develops within the body of the mother, and is fed by milk produced by the mammary glands. The mammals are warm-blooded and the body is covered with hair. In Mammals, female is the heterogametic sex (XY) and thus male-to-male competition is the predominant form of sexual selection.

Manx. Tailless phenotype in cats, caused by an autosomal dominant mutation that is lethal when homozygous.

map unit (m.u.). The distance between two linked gene pairs where 1 percent of the products of meiosis are recombinant. A unit of distance in a linkage map. The distance equal to 1% recombination between two loci.

mapping function. The mathematical relationship between measured map distance and actual recombination frequency. A formula expressing the relationship between distance in a linkage map and recombinant frequency.

mapping. The study of the position of genes on chromosomes.

Marfan's syndrome. A human disorder of the connective tissue expressed as a range of symptoms, including very long limbs and digits, and heart defects; inherited as an autosomal dominant phenotype. Mutations in a gene called fibrillin.

marker retention. A technique used in yeast to test the degree of linkage between two mitochondrial mutations.

marker. (1) Any genetic element (locus, allele, DNA sequence or chromosome feature) which can be readily detected by phenotype, cytological or molecular techniques, and used to follow a chromosome or chromosomal segment during genetic analysis. See centromere marker; chromosome marker; DNA marker; genetic marker; inside marker; outside marker. (2) Two typical usages:

Molecular weight size marker: a piece of DNA of known size, or a mixture of pieces with known size, used on electrophoresis gels to determine the size of unknown DNA's by comparison.

Genetic marker: A known site on the chromosome. It might for example be the site of a locus with some recognizable phenotype, or it may be the site of a polymorphism that can be experimentally discerned.

mate-killer. A phenotype of Paramecium induced by intracellular bacteria-like mu particles.

maternal effect lethal. A recessive mutation that is viable in zygotes, but in which homozygous mutant mothers produce inviable offspring.

maternal effect lethals. One form of selfish/parasitic DNA that facilitates its own propagation. They are post-zygotic distorters that kill progeny lacking the factor. *Medea* in beetles and *Scat* in mice are the known examples. Progeny of the heterozygous mothers that are homozygous for wild-type are killed. Progeny carrying a copy of the lethal are actually protected.

maternal effect. The effect of the maternal parent's genotype on the phenotype of her offspring.

maternal inheritance. (1) A type of uniparental inheritance in which all progeny have the genotype and phenotype of the female parent. (2) Diseases due to mutations in mtDNA are transmitted only by mothers because all mitochondria are inherited via the egg. Thus, all offspring of an affected female are at risk of inheriting the abnormality, whereas no offspring of an affected male are at risk. Clinical manifestations are variable and may be due to variable mixtures of mutant and normal mitochondrial genomes (heteroplasmy) within cells and tissues.

mating type. (1) Genetically determined characteristics of bacteria, ciliates, fungi and algae, determining their ability to conjugate and undergo sexual reproduction with other members of the species. In yeasts (S.cerevisiae) which have only two types, only cells of opposite types can conjugate. The common mushroom

Schizophyllum has more than 50,000 mating types (genders) encoded in two separate loci. In species where the organelles are inherited uniparentally and reproduction is by the union of gametes, there are only two mating types. In species reproducing via sexual conjugation (nuclear exchange) so that each cell preserves its own organelles, there can be multiple types. (2) In many species of microorganisms individuals can be divided into two types. Mating can take place only between individuals of opposite mating types due to the interaction of cell surface components. The equivalent in lower organisms of the sexes in higher organisms; the mating types typically differ only physiologically and not in physical form.

matroclinous inheritance. Inheritance in which all offspring have the phenotype of the mother.

mean fitness of the population W(bar). The sum of the fitnesses of the genotypes of a population weighted by their proportions; hence a weighted mean fitness.

mean. The arithmetic average; the sum of the data divided by the sample size.

medium. Any material on (or in) which experimental cultures are grown.

meiocyte. Cell in which meiosis occurs.

meiosis. (1) Cell division with two phases resulting four haploid cells (gametes) from a diploid cell. In meiosis I, the already doubled chromosome number reduces to half to create two diploid cells each containing one set of replicated chromosomes. Genetic recombination between homologous chromosome pairs occurs during meiosis I. In meiosis II, each diploid cell creates two haploid cells resulting in four gametes from one diploid cell. (2) Two successive nuclear divisions (with corresponding cell divisions) that produce haploid gametes (in animals) or haploid sexual spores (in plants and fungi) having one-half of the genetic material of the original cell. The nuclear and cell division process in diploid eukaryotes that results in four haploid gametes or spores having one member of each original pair of homologous chromosomes only per nucleus.

meiospore. Cell that is one of the products of meiosis in plants.

melting of DNA. Denaturation of duplex DNA by heat or increased pH leading to strand separation.

melting temperature (*Tm*)**.** The temperature at which the two strands of a double-stranded DNA molecule come apart. A short (<18 nucleotides) oligonucleotides *Tm* value (^{0}C) is estimated by the formula: *Tm* = (number of A + T)x2 + (number of G + C)x4.

Mendelian inheritance. Inheritance of traits mediated by nuclear genes (as opposed to mitochondrial DNA) according to the laws defined by Gregor Mendel.

Mendelian ratio. A ratio of progeny phenotypes reflecting the operation of Mendel's laws.

Mendel's first law (law of segregation). (1) The two alleles received one from each parent segregate independently in gamete formation, so that each gamete receives one or the other with equal probability. (2) The two members of a gene pair segregate from each other during meiosis; each gamete has an equal probability of obtaining either member of the gene pair.

Mendel's second law (law of recombination). (1) Two characters determined by two unlinked genes are recombined at random in gametic formation, so that they segregate independently of each other, each according to the first law (note that recombination here is not used to mean crossing-over in meiosis). (2) The law of independent assortment; unlinked or distantly linked segregating gene pairs assort independently at meiosis.

merozygote. A bacterial cell having a second copy of a particular chromosomal region in the form of an exogenote. A partially diploid Escherichia coli cell formed from a complete chromosome (the endogenote) plus a fragment (the exogenote).

message. See mRNA.

messenger RNA (mRNA). An RNA molecule transcribed from the DNA of a gene, and from which a protein is translated by the action of ribosomes. The basic function of the nucleotide sequence of mRNA is to determine the amino acid sequence in proteins.

Met. Methionine (an amino acid).

metabolism. The chemical reactions that occur in a living cell.

metacentric chromosome. (1) A chromosome having its centromere in the middle. See also acrocentric chromosome. (2) A chromosome with its centromere near the center. If the centromere is slightly off-center, the chromosome is said to be submetacentric (see also acrocentric and telocentric).

metafemale. A fruit fly with an X chromosome:Autosome ratio greater than unity.

metagon. An RNA necessary for the maintenance of mu particles in Paramecium.

metamale. A fruit fly with X chromosome:Autosome ratio below 0.5.

metamere. Segmental repeat unit in higher animals.

metaphase plate. The plane of the equator of the spindle into which chromosomes are positioned during metaphase.

metaphase. (1) Mitotic phase at which replicated chromosomes are fully condensed and become visible under the light microscope. (2) The stage of mitosis or meiosis in which spindle fibers are attached to kinetochores and the chromosomes are positioned in the equatorial plane of the cell. An intermediate stage of nuclear division when chromosomes align along the equatorial plane of the cell spindle.

metastasis. The migration of cancerous cells to other parts of the body.

metazoa. A major division in the Animal Kingdom consisting of multicellular animals.

methylation paradox. Methylation of the CpG islands in the transcribed region is often correlated with transcription but an inverse correlation is seen at the CpG islands at the transcription initiation site.

methylation. (1) Modification of a molecule by the addition of a methyl group. (2) The addition of a methyl group (-CH_3) to DNA. Methylated DNA is inactivated and not transcribed. Most frequently occurs at CpG doublets (see genomic imprinting and CpG islands).

metrical variation. See continuous variation.

MHC restriction. The phenomenon that a T cell can only recognize a peptide when it is (processed and) presented by another cell sharing the same MHC type. The only exception is that a foreign MHC antigen itself does not have to be presented by a cell but is able to induce a reaction directly (as happens in MHC - mismatched transplantation).

MHC. Major Histocompatibility Complex. H-2 complex in mice, HLA complex in humans.

microlymphocytotoxicity (MLC) assay. An assay used in the typing of HLA molecules (serological typing).

microsatellite repeat sequences. Sequences of 2 to 5 bp repeated up to 50 times such as a TA dinucleotide repeat polymorphism. The variable number of repeats creates the polymorphism. They may occur at 50 - 100 thousand locations in the human genome. Microsatellites mutate faster than nonrepeat polymorphisms and can be used to estimate evolutionary relationships over shorter time scales.

microtubules. Hollow cylinders made of the protein **tubulin** that form, among other things, the spindle fibers.

midparent value. The mean of the values of a quantitative phenotype for two specific parents.

migration. Movement of individuals between otherwise reproductively isolated populations.

mimicry. (1) A phenomenon in which an individual; gains an advantage by looking like the individuals of a different species. (2) The resemblance of one kind of organism to another to make the organism difficult to find, to discourage the potential predators, or to attract potential prey. The common kinds of mimicry are Batesian and Mullerian mimicry. See also molecular mimicry.

minimal medium. A culture medium for microorganisms that contains the minimal necessities for growth of the wild-type. A medium containing only inorganic salts, a carbon source, and water.

mismatch repair. A form of excision repair initiated at the sites of mismatched bases in DNA.

missense mutation. (1) A mutation that causes the substitution of one amino acid for another (non-synonymous change). An example is the major HFE mutation C282Y in which results in an amino acid change at position 282. (2) Mutations that at change a codon for one amino acid into a codon for a different amino acid.

missing link. An absent member needed to complete an evolutionary lineage.

mitochondrial cytopathy. Human disorder caused by point mutation or deletion in mitochondrial DNA; inherited maternally.

mitochondrial DNA. The maternally inherited nucleic acid found in cytoplasm whose homologue in plants is chloroplastic DNA. This small circular DNA codes for tRNAs, rRNAs and some mitochondrial DNAs. It is more closely related to bacterial DNA than to eukaryotic nuclear DNA. Mitochondrial DNA mutates 10-20 times faster than nuclear DNA.

mitochondrion. A eukaryotic organelle that is the site of ATP synthesis and of the citric acid cycle. The eukaryotic cellular organelle in which the Krebs cycle and electron transport reactions take place.

mitosis. Cell division into two identical daughter cells with the same chromosome number as the mother cell (see also meiosis). Replicated chromosomes separate and each chromatid goes to a daughter cell.

mitosis. The nuclear division producing two daughter nuclei identical to the original nucleus. A type of nuclear division that produces two daughter nuclei identical to the parent nucleus normally just prior to cell division. See endomitosis, cytokinesis.

mitotic apparatus. See spindle.

mitotic crossover. A crossover resulting from the pairing of homologous chromosomes in a diploid during mitosis.

mitotic recombination. During mitosis sister chromatids freely exchange material without changing anything in genetic material because they are identical. Very rarely, and by chance, homologous chromosomes come very close to each other and exchange material as in **meiosis** which results in a recombinant chromosome.

mixed codon family. Group of four codons sharing their first two bases but coding for more than one amino acid.

mixed lymphocyte reaction (MLR). The activation of T cells *in vitro* by other (allogeneic) lymphocytes due to differences in their MHC molecules.

mobile genetic element. See transposable genetic element.

mode. The single class in a statistical distribution having the greatest frequency.

modifier gene. A gene that affects the phenotypic expression of another gene.

molecular chaperone. A protein that aids in the folding of a second protein. The chaperone prevents proteins from taking conformations that would be inactive.

molecular evolutionary clock. A measurement of evolutionary time in nucleotide substitutions per year.

molecular genetics. The study of the molecular processes underlying gene structure and function.

molecular imprinting. The phenomenon in which there is differential expression of a gene depending on whether it was maternally or paternally inherited. Paternal imprinting means that an allele inherited from the father is not expressed in offspring. Maternal imprinting means that an allele inherited from the mother is not expressed in offspring.

molecular mimicry. Resemblance of a DNA sequence or a polypeptide by an unrelated sequence at the nucleotide or amino acid level, respectively. Mimicry of MHC proteins is an immunoevasion mechanism used by pathogens.

monera. One of the five Kingdoms which contains all prokaryotes. It contains archaebacteria, eubacteria and cyanobacteria. The first life form emerged over 3,500 Mya were the members of this Kingdom from which eukaryotes evolved.

monocistronic mRNA. An mRNA that codes for only one protein.

monoclonal antibody. The unique immunoglobulin molecule (antibody) produced by a clone of cells derived from a fusion of a B lymphocyte with a myeloma cell.

monoecious plant. A plant species in which male and female organs are found on the same plant but in different flowers (for example maize).

monohybrid cross. A cross between two individuals identically heterozygous at one gene pair for example, Aa x Aa.

monohybrid. Offspring of parents that differ in only one genetic characteristic. Usually implies heterozygosity at a single locus under study.

monoploid. A cell having only one chromosome set (usually as an aberration) or an organism composed of such cells.

monosomic. A diploid cell missing a single chromosome. A cell or individual that is basically diploid but that has only one copy of one particular chromosome type and thus has chromosome number 2n - 1.

monotreme. A subclass of the Class Mammalia consisting of animals that lay eggs.

morphological species concept. Organisms are classified in the same species if they appear identical by morphological (anatomical) criteria.

mosaic. A chimera; a tissue containing two or more genetically distinct cell types, or an individual composed of such tissues. Individual made up of two or more genetically distinct cell lines.

mosaicism. (1) Mosaicism is the presence of more than one cell lines differing in genotype or karyotype but derived from one zygote. Post-zygotic new mutations result in mosaic individuals who may not be clinically affected themselves, but are at risk of bearing multiple affected offspring. Mosaicism is well recognized in Duchenne muscular dystrophy and in autosomal dominant disorders with high new mutation rates. (2) The condition of being a mosaic. See mosaic.

mRNA. (1) "Messenger RNA" or sometimes just "message"; an RNA which contains sequences coding for a protein. The term mRNA is used only for a mature transcript with polyA tail and with all introns removed, rather than the primary transcript in the nucleus. As such, an mRNA will have a 5' untranslated region, a coding region, a 3' untranslated region and (almost always) a poly(A) tail. Typically about 2% of the total cellular RNA is mRNA. (2) It is the first product of the DNA transcription by RNA polymerase. mRNA forms 1-5% of the total cellular RNA. Its molecular weight is generally less than $2x10^6$. At any time, there are about 10^5 species of mRNA in a cell.

mtDNA. Mitochondrial DNA.

mu particle. Bacteria-like particle found in the cytoplasm of Paramecium that have the mate-killer phenotype.

mu phage. A kind (species) of phage with properties similar to those of insertion sequences, being able to insert, transpose, and cause chromosome rearrangements and mutations.

Muller's ratchet. The continual decrease in fitness due to accumulation of (usually deleterious) mutations without compensating mutations and recombination in an asexual lineage. Recombination (sexual reproduction) is much more common than mutation, so it can take care of mutations as they arise. This is one of the reasons why sex is believed to have evolved.

Mullerian mimicry. A form of mimicry in which noxious species evolve to resemble each other.

multihybrid. An organism heterozygous at numerous loci.

multimeric structure. A structure composed of several identical or different subunits held together by weak bonds.

multinomial expansion. The terms generated when a multinomial is raised to a power.

multiple allelism. The existence of several known alleles of a gene.

multiple-factor hypothesis. A hypothesis to explain quantitative variation by assuming the interaction of a large number of genes (polygenes) each with a small additive effect on the character.

multiple-hit hypothesis. The proposal that a single cell must receive a series of mutational events in order to become malignant or cancerous.

multiplicity of infection (moi). The average number of phage particles that infect a single bacterial cell in a specific experiment.

multivariate analysis. A statistical analysis of several variables assessed simultaneously. Should be preferred over individual analyses of pairs of variables.

mutability. The ability to change.

mutagen. An agent that is capable of increasing the mutation rate .

mutant allele. An allele differing from the allele found in the standard or wild type organism.

mutant screening. The process of screening for mutations showing abnormalities in a certain structure or in a particular function, as a preparation for genetic analysis of that strucutre or function.

mutant site. The damaged or altered area within a mutated gene.

mutant. An organism or cell carrying a mutation. An alternative phenotype to the wild-type; the phenotype produced by a non-wildtype allele.

mutation breeding. Use of mutagens to develop variants that can increase agricultural yield.

mutation event. The actual occurrence of a mutation in time and space.

mutation frequency. The frequency of mutations in a population.

mutation pressure. Evolution by different mutation rates alone.

mutation rate. (1) The number of mutation events per gene per unit of time (for example, per cell generation). The proportion of mutations per cell division in bacteria or single-celled organisms or the proportion of mutations per gamete in higher organisms. (2) The number of mutations at a particular locus which occur per gene per cell generation. This is the only source of variation in asexual organisms. The mutation rate is the likelihood of parentage when findings suggest otherwise. Beware of the different units in different mutation rates. In humans, the mutation rate is 1 bp per 10^9 bp per cell division. This corresponds to 10^{-6} mutations per gene per cell division and because there are 10^{16} divisions in a lifetime, 10^{10} mutations per gene per lifetime.

mutation. (1) The process producing a gene or a chromosome differing from the wild-type. (2) The gene or chromosome that results from such a process. (3) Any heritable change (not only point mutation) brought about by an alteration in the genetic material. Includes gene conversion, deletion, .duplication, insertion and so forth.

mutational dissection. The study of the components of a biological system through a study of mutations affecting that system.

mutational load. Genetic load caused by mutation.

mutator mutation. mutation of DNA polymerase that increases the overall mutation rate.

muton. A term coined by Benzer for the smallest mutable site within a cistron. The smallest part of a gene that can be involved in a mutation event; now known to be a single nucleotide pair.

Mya. Million years ago.

myeloma. A cancer of the bone marrow.

N

N segment. Sequence of nucleotides added in a template-independant fashion at the joining junctions of heavy-chain immunoglobulin genes.

narrow heritability. See heritability in the narrow sense.

natural killer (NK) cell. Bone marrow-derived, mononuclear white blood cells (large granular lymphocytes) that are able to kill invading microorganisms without activation by cells of the immune system. They are, therefore, part of the innate immune system. They are specialized in killing virus-infected cells and cells transformed to develop cancer.

natural selection (Darwin's definition, 1859). "As many more individuals of each species are born than can possibly survive; and as, consequently, there is a frequent recurrent struggle for existence, it follows that any being, if it varies however slightly in any manner profitable to itself, under the complex and sometimes varying conditions of life, will have a better chance of surviving, and thus be naturally selected."

natural selection. The process in nature whereby one genotype leaves more offspring than another genotype because of superior life history attributes (fitness) such as survival or fecundity.

nature-nurture debate. The debate on the relative contributions of genetics (nature) and environment (nurture) to the characteristics of an organism. An example is the debate on whether gene(s) and/or environmental factors determine the sexual orientation of an individual. Finding a gene playing a role in the development of a condition does not necessarily mean it is a purely genetic trait.

nearest-neighbour analysis. A technique of transferring radioactive atoms between adjacent nucleotides in DNA used to demonstrate that the two strands of DNA run in opposite directions.

negative assortative mating. (1) A type of nonrandom mating in which individuals of unlike phenotype mate more often than predicted under random mating conditions. (2) Preferential mating between phenotypically different partners.

negative control. Transcription regulation mediated by factors that block, turn down or turn off transcription.

negative interference. The phenomenon whereby a crossover in a particular region enhances the occurrence of other apparent crossovers in the same region of the chromosome.

N-end rule. The life span of a protein is determined by its amino-terminal (N-terminal) amino acid.

neo-darwinism. The merger of classical Darwinian evolution with population genetics.

neomorph. Mutant showing a novel substance or structure not found in wild-types.

neoplasm. New growth of abnormal tissue.

neoteny. Retention of juvenile features in sexually mature adult animals. Neoteny frequently correlates with recent evolution of the species (like Homo sapiens).

neurofibromatosis. A human disease with tumours of nerve cells all over the body and pigmented spots in the skin. The allele generally arises from germinal mutation, but it is inherited as an autosomal dominant.

neurospora crassa. Haploid, heterothallic, filamentous Ascomycete fungus (bread mold). It has two mating types (A and a) operating as sexual compatibility system, and 11 het loci operating as heterokaryon compatibility system in vegetative phase.

Neurospora. A pink mold, commonly found growing on old food.

neutral gene hypothesis. The hypothesis that most genetic variation in natural populations is not maintained by selection because most alleles have equal fitness.

neutral mutation. (1) A mutation that has no effect on the Darwinian fitness of its carriers. (2) A mutation that has no phenotypic effect.

neutral petite. A petite yeast that produces all wild-type progeny when crossed with wild-type.

neutrality. See selective neutrality.

NF. See fundamental number.

nick translation. A method for incorporating radioactive isotopes (typically 32P) into a piece of DNA. The DNA is randomly nicked by DNase I, and then starting from those nicks DNA polymerase I digests and then replaces a stretch of DNA. Radiolabeled precursor nucleotide triphosphates can thus be incorporated.

nickase. See DNA gyrase.

nicking. Nuclease action to sever the sugar-phosphate backbone in one DNA strand but not the other at one specific site.

nitrocellulose filter. A type of filter used to bind DNA for hybridization.

nitrogen base. Type of molecule that forms an important part of nucleic acid, composed of a nitrogen-containing ring structure. Hydrogen bonds between bases in opposing complementary strands link the two strands of a DNA double helix.

non-coding region. Parts of a gene that include sequences which are not translated. Both 5' and 3' untranslated regions (UTRs), upstream promoter region and introns are classified as non-coding regions.

non-coding strand. (1) Anti-sense strand. See "Sense strand" for a discussion of sense strand vs. anti-sense strand. (2) See anticoding strand.

non-disjunction. (1) Due to failure in pairing of homologous chromosomes in meiosis, the two members of one pair migrate to the same pole, giving rise to unbalanced gametes, one of which contain both homologous chromosomes, and the other none (most frequent in sex chromosomes). The non-disjunction event is much more frequent in maternal meiosis I. This may be due to the fact that in a mature woman, oocytes have been held in the ovary for a very long time at prophase I of meiosis from before her birth to shortly before ovulation of the oocyte in question. (2) The failure of a pair of homologous chromosomes to separate properly during meiosis. The failure of homologues (at meiosis) or sister chromatids (at mitosis) to separate properly to opposite poles, that is two chromosomes or chromatids go to one pole and none to the other.

non-histone protein. The protein remaining in chromatin after the histones are removed. The scaffold structure is made of nonhistone proteins.

non-linear tetrad. A tetrad in which the meiotic products are in no particular order.

non-Mendelian ratio. An unusual ratio of progeny phenotypes that does not reflect the simple operation of Mendel's laws; for example, mutant:wild-type ratios of 3:5, 5:3, 6:2, or 2:6 in tetrads indicate that gene conversion has occurred.

non-parental ditype (NPD). A spore arrangement in Ascomycetes that contains only the two recombinant-type ascospores (assuming two segregating loci). A tetrad type containing two different genotypes, both of which are recombinant.

non-parental. See recombinants.

non-recombinant. In mapping studies the offspring that have alleles arranged as in the original parents are non-recombinants.

nonsense codon. One of the mRNA sequences (UAA, UAG, UGA) that signals the termination of translation. A codon for which no normal tRNA molecule exists. The presence of a nonsense codon causes termination of translation (ending polypeptide chain synthesis). There are three nonsense codons are called amber(UAG) ochre(UAA) and opal(UGA).

nonsense mutation. A mutation that changes an amino acid specifying codon to one of the three termination (stop) codons.

nonsense suppressor. A mutation in the anticodon of tRNA that alters the anticodon so it is now complementary to a nonsense codon allowing the tRNA to insert its cognate amino acid at this nonsense codon during translation.

norm of reaction. The pattern of phenotypes produced by a given genotype under different environmental conditions.

normal distribution. Any of a family of bell-shaped frequency curves whose relative position and shape are defined on the basis of the mean and standard deviation.

northern blot. (1) A technique for analyzing mixtures of RNA, whereby the presence and rough size of one particular type of RNA (usually an mRNA) can be ascertained. See "Blotting" for more information. After Dr. E. M. Southern invented the Southern blot, it was adapted to RNA and named the "Northern" blot. (2) Transfer of electrophoretically separated RNA molecules from a gel onto an absorbent sheet, which is then immersed in a labeled probe that will hybridize to an RNA of interest to reveal its presence. See Southern blotting.

notochord. A rod that forms in the embryonic mesoderm and which establishes the front-to-back orientation of vertebrate embryos. It also initiates the formation of the nervous system, the skeleton and most muscles.

NPD. See nonparental ditype.

NT. Abbreviation for nucleotide; i.e. the monomeric unit from which DNA or RNA are built. One can express the size of a nucleic acid strand in terms of the number of nucleotides in its chain; hence 'nt' can be a measure of chain length.

nu body. See nucleosome.

nuclear run-on. A method used to estimate the relative rate of transcription of a given gene, as opposed to the steady-state level of the mRNA transcript (which is influenced not just by transcription rates, but by the stability of the RNA). This technique is based on the assumption that a highly-transcribed gene should have more molecules of RNA polymerase bound to it than will the same gene in a less-active state. If properly prepared, isolated nuclei will continue to transcribe genes and incorporate 32P into RNA, but only in those transcripts that were in progress at the time the nuclei were isolated. Once the polymerase molecules complete the transcript they have in progress, they should not be able to re-initiate transcription. If that is true, then the amount of radiolabel incorporated into a specific type of mRNA is theoretically proportional to the number of RNA polymerase complexes present on that gene at the time of isolation. A very difficult technique, rarely applied appropriately from what I understand.

nuclear transplantation. The technique of placing a nucleus from another source into an enucleated cell.

nuclease hypersensitive site. Region of eukaryotic chromosome DNA that is specifically vulnerable to nuclease attack perhaps because it is not wrapped in histone as nucleosomes.

nuclease protection assay. See "RNase protection assay".

nuclease. (1) An enzyme that breaks bonds in nucleic acids. Deoxyribonuclease (DNAase) and ribonuclease (RNAase). (2) An enzyme which degrades nucleic acids. A nuclease can be DNA-specific (a DNase), RNA-specific (RNase) or non-specific. It may act only on single stranded nucleic acids, or only on double-stranded nucleic acids, or it may be non-specific with respect to strandedness. A nuclease may degrade only from an end (an exonuclease), or may be able to start in the middle of a strand (an endonuclease). To further complicate matters, many enzymes have multiple functions; for example, Bal31 has a 3'-exonuclease activity on double-stranded DNA, and an endonuclease activity specific for single-stranded DNA or RNA. (3) One of the several classes of enzymes that degrade nucleic acid. An enzyme that can degrade DNA or RNA by breaking phosphodiester bonds. See endonucleases; exonucleases.

nucleoid. (1) A DNA mass within a chloroplast or mitochondrion. (2) The loosely tangled clump of DNA within the cytoplasm of a prokaryotic cell.

nucleolar organizer. (1) A region on a chromosome that is associated with formation of a new nucleolus following cell division. It contains the genes for several species of ribosomal RNA (rRNA), i.e., 18S, 5.8S, 5S and 28S in eukaryotes. (2) The chromosomal region around which the nucleolus forms, a site of tandem repeats of the rRNA gene. A region (or regions) of the chromosome set physically associated with the nucleolus and containing rRNA genes.

nucleolus. (1) The globular nuclear organelle formed at the nucleolar organizer. Site of ribosome construction. An organelle found in the nucleus, containing rRNA and multiple copies of the genes coding for rRNA. (2) The site of synthesis of rRNA within the nucleus of a eukaryotic cell.

nucleoprotein. The substance of eukaryotic chromosomes consisting of proteins and nucleic acids.

nucleoside. (1) A small molecule composed of a purine or pyrimidine base linked to a five-carbon sugar (pentose: ribose or deoxyribose). With the addition of a phosphate group, it becomes a nucleotide. Nucleosides in RNA are adenosine, guanosine, cytidine and uridine; in DNA, they are (d)adenosine, (d)guanosine, (d)cytidine and (d)thymidine. (2) A sugar-base compound that is a nucleotide precursor. Nucleotides are nucleoside phosphates. A nitrogen base linked to a sugar molecule.

nucleosome. (1) A beadlike structure of eukaryotic chromosomes. It consists of a core of eight histone molecules and a DNA segment of about 150 base pairs. Each nucleosome is separated from another by a linker DNA sequence of about 50 base pairs. Nucleosome structure helps to fold DNA into a compact form in the interphase nucleus. Otherwise the length of a chromosome, when linear, is many orders of magnitude greater than the diameter of the nucleus. (2) A nu body; the basic unit of eukaryotic chromosome structure; a ball of eight histone molecules wrapped about by two coils of about 220 base pairs of DNA. Arrangement of DNA and histones forming regular spherical structures in eukaryotic chromatin.

nucleotide pair. A pair of complementary nucleotides (one in each strand of DNA) that are joined by hydrogen bonds.

nucleotide. (1) Subunit that polymerizes into nucleic acids (DNA or RNA). Each nucleotide consists of a nitrogenous base; a sugar; and one to three phosphate groups. (2) The monomeric unit that makes up the DNA or RNA, formed by a phosphate group, a pentose and one of the nitrogenous bases (A, T/U, C, G). Nucleotides in RNA are adenylate, guanylate, cytidylate and uridylate; in DNA, they are (d)adenylate, (d)guanylate, (d)cytidylate and thymidylate.

nucleotide-pair substitution. The replacement of a specific nucleotide pair by a different pair; often mutagenic.

nucleus. The membrane bound organelle containing the genome of eukaryotes organized into chromosomes.

null allele. An allele whose effect is either an absence of normal gene product at the molecular level or an absence of normal function at the phenotypic level.

null hypothesis. The statistical hypothesis that states that there are no differences between observed and expected data.

nullisomic. A diploid cell or individual missing both copies of the same chromosome. A cell or individual with one chromosomal type missing, with a chromosome number such as n-1 or 2n-2.

numerator element. Genes on the X chromosome in Drosophila that regulate the sex switch (sxl) to the **on** condition (femaleness). Refers to the numerator of the X/A (number of X chromosomes/autosomes) genic balance equation.

nurse cell. A sister cell of an oocyte in insects. Nurse cells produce the bulk of the cytoplasmic contents of the mature oocyte.

nutritional-requirement mutant. See auxotroph.

O

ochre codon. The codon UAA, a nonsense codon.

octad. An ascus containing eight ascospores, produced in species in which the tetrad normally undergoes a postmeiotic mitotic division.

Okazaki fragment. (1) Segment of newly replicated DNA produced during discontinuous DNA replication. (2) The small (<1 kb), discontinuous strands of DNA produced using the lagging strand as template during DNA synthesis. DNA ligase links the Okazaki fragments to give rise to a continuous strand.

olfactory. Related to smell.

oligonucleotide ligation assay (OLA). A PCR-based method for SNP typing. It is a ligase mediated gene detection system which uses exact 3' matching of a primer to one of the SNP allele. If this happens the other labeled oligonucleotide which binds to the nucleotide immediately next to the SNP on the other side would be joined to the primer by ligase. The resulting sample can then be tested for the presence of the label (for example biotin). Unless controls are included, false positives are possible.

oligonucleotide. A short, synthetic DNA string used as a probe (as in SSOP) or primer (as in SSP) in molecular genetic studies.

oncogene. (1) A gene capable of causing malignant transformation. (2) A gene in a tumor virus or in cancerous cells which, when transferred into other cells, can cause transformation (note that only certain cells are susceptible to transformation by any one oncogene). Functional oncogenes are not present in normal cells. A normal cell has many "proto-oncogenes" which serve normal functions, and which under the right circumstances can be activated to become oncogenes. The prefix "v-" indicates that a gene is derived from a virus, and is generally an oncogene (like *v-src*, *v-ras*, *v-myb* , etc). See also "Transformation (with respect to cultured cells)". (3) A gene that contributes to the production of a cancer. Oncogenes are generally mutated forms of normal cellular genes (proto-oncogenes). A gene capable, when activated, of transforming a cell. Oncogenes are found in the oncogenically activated state in retroviruses and transformed cells and in their normal non-oncogenically activated state in non-transformed cells in which they are called proto-oncogenes.

one gene-one (enzyme) polypeptide hypothesis. The hypothesis that each gene controls the synthesis of a single polypeptide which may be a subunit of a complex protein.

one-gene-one-enzyme hypothesis. Hypothesis of Beadle and Tatum that one gene controls the production of one enzyme. Later modified to the concept that one cistron controls the production of one polypeptide.

oocyte. (1) Female sex cell which undergoes meiosis and produces an egg (ovum). (2) The gamete in females. See ovum.

oogenesis. The process of ovum formation in female animals.

oogonia. Cells in females that produce primary oocytes by mitosis.

opal codon. The codon UGA, a nonsense codon.

open reading frame (ORF). (1) A nucleotide sequence encoding a polypeptide starting with a start and ending with a stop codon. (2) Any region of DNA or RNA where a protein could be encoded. In other words, there must be a string of nucleotides (possibly starting with a Met codon) in which one of the three reading frames has no stop codons. See "Reading frame" for a simple example.

operator. A DNA region at one end of an operon that acts as the binding site for repressor protein. A DNA sequence that is recognized by a repressor protein or repressor-corepressor complex. When the operator is complexed with the repressor, transcription is prevented.

operon. (1) A set of adjacent structural genes in bacteria whose mRNA is synthesized in one piece, together with the adjacent regulatory signals that affect transcription of the structural genes. A sequence of adjacent bacterial genes all under the transcriptional control of the same operator. (2) A type of genetic unit which consists of one or more transcription units that are transcribed together into a polycistronic mRNA. The transcription of each operon is initiated at a promoter region and controlled by a neighboring regulatory gene (an operator which binds to a repressor or an apoinducer, to repress or induce the transcription, respectively). An example is the *lac* operon of E.coli.

ORF (open reading frame). A section of a sequenced piece of DNA that begins with an initiation (methionine ATG) codon and ends with a nonsense codon. ORFs all have the potential to encode a protein or polypeptide, however many may not actually do so.

organelle. A subcellular structure having a specialized function for example the mitochondrion, the chloroplast, or the spindle apparatus.

organogenesis. The production of organ systems during animal embryogenesis.

ori. The origin of replication in prokaryotes.

origin of replication. (1) Nucleotide sequences present in a plasmid which are necessary for that plasmid to replicate in the bacterial host. (Abbr. "ori"). (2) The point of specific sequence at which DNA replication is initiated.

orthology. Being homologous by descent. In other words, descendants from a common ancestor. An example is the MHC class II genes in different species that all descended from a common ancestral class I gene.

outbreeding. The mating of genetically unrelated individuals.

outside markers. Loci on either side of another locus or specified region.

overdominance. A relationship in which the phenotypic expression of the heterozygote is greater than that of either homozygote.

overdominance. See **balancing selection**. See also **underdominance**.

oviparity. Egg birth as opposed to live birth (viviparity).

ovum Egg. The one functional product of each meiosis in female animals.

P

P (peptidyl) site. The site on the ribosome occupied by the peptidyl-tRNA just before peptide bond formation.

P element. A **Drosophila** transposable element that has been used as a tool for insertion mutagenesis and for germline transformation.

P1. Parental generation.

P53. A 53KDa protein the product of a tumour suppressor gene. The loss of this protein due to mutation is a primary event in the formation of many types of cancer (breast, colon, lung, leukaemia, liver). P53 is involved in regulating the activity of some other genes (e.g. P21, a cell division kinase inhibitor). It also prevents cells entering cell division. P53 levels are increased after DNA is damaged by UV and ionizing radiation and cells are arrested in cell division until either the damage is repaired or they die by apoptosis (programmed cell death).

pachynema (pachytene stage). The stage of prophase 1 of meiosis in which chromatids are first distinctly visible.

paleontology. The study of fossils.

palindrome. A sequence of letters, words, phrases, or nucleotides that reads the same regardless of direction, for instance what Adam said to Eve "Madam I'm

Adam". The DNA sequence of type II restriction endonuclease recognition sites (e.g., EcoRI 5'GAATTC3') and the binding sites of many other DNA binding proteins (e.g. catabolite activator protein (CAP) binding site 5'GTGAGTTAGCTCAC3') are not true palindromes but in double stranded form are similar and have two-fold rotational symetry.

panmictic. Referring to unstructured (random-mating) populations.

paracentric inversion. An inversion not involving the centromere. A chromosomal inversion that does not include the centromere.

parallel evolution. Evolution of roughly similar changes in two or more closely related lineages.

paralogous genes. Two genes or clusters of genes at different chromosomal locations in the same organism that have structural similarities indicating that they derived from a common ancestral gene.

paralogy. Being homologous due to a recent or past duplication in the same species. An example is the chromosomal regions 1q22-q23; 6p21.31; and 9q33-q34 in humans. These regions contain very similar genes including some of the MHC class III genes.

paramecin. A toxin liberated by a killer Paramecium.

paramecium. A unicellular Protoctist belonging to the group Ciliates. Although normally reproduces asexually, they also undergo sexual conjugation in which mating types play a role. Paramecium aurelia has 34 hereditary mating types that form 16 distinct mating groups.

parameters. Measurements of attributes of a population.

paramutation. In paramutation, two alleles of a gene interact so that one of the alleles is epigenetically silenced. The silenced state is then genetically transmissible for many generations.

parapatric speciation. (1) Speciation in which the evolution of reproductive isolating mechanisms occurs when a population enters a new niche or habitat within the range of the parent species. (2) Speciation that occurs as a result of two populations diverging in adjacent geographical areas.

parental ditype (PD). A tetrad type containing two different genotypes, both of which are parental. A spore arrangement in Ascomycetes that contains only the two non-recombinant-type ascospores.

parental imprinting. See molecular imprinting.

parental. See non-recombinant.

parsimony. The scientific convention whereby the simplest explanation is preferred over the others. This is usually a phylogenetic tree requiring the fewest evolutionary steps.

parthenogenesis (virgin birth). Reproduction involving unfertilized eggs. The offspring of parthenogenetic parents are less diverse than those of sexual parents. Parthenogenesis often alternates with sexual reproduction. Two examples are the desert grassland whiptail lizard (Cnemidophorus uniparens) and Caucasian rock lizard. Never seen in birds and mammals.

parthenogenesis. The production of offspring by a female with no genetic contribution from a male. The development of an individual from an unfertilized egg that did not arise by meiotic chromosome reduction .

partial digest. A restriction digest that has not been allowed to go to completion and thus contains pieces of DNA with some restriction endonuclease sites that have not yet been cleaved.

partial diploid. See merozygote.

partial dominance. See incomplete dominance.

particulate inheritance. The model proposing that genetic information is transmitted from one generation to the next in discrete units (particles) so that the character of the offspring is not a smooth blend of essences from the parents (See blending inheritance).

Pascal's triangle. A triangular array of numbers made up of the coefficients of the binomial expansion.

path diagram. A modified pedigree showing only the direct line of descent from common ancestors.

pathogen. An organism that causes disease in another organism.

patroclinous inheritance. Inheritance in which all offspring have the nucleus-based phenotype of the father.

pattern formation. The developmental processes by which the complex shape and structure of higher organisms occurs.

pBR322. A common plasmid. Along with the obligatory origin of replication, this plasmid has genes which make the *E. coli* host resistant to ampicillin and tetracycline. It also has several restriction sites (BamHI, PstI, EcoRI, HindIII etc.) into which DNA fragments could be spliced in order to clone them.

PCR. Polymerase chain reaction. A technique that allows amplification of specific DNA segments in a very short time.

PD. See parental ditype.

pedigree. A family tree drawn with standard genetic symbols, showing inheritance patterns for specific phenotypic characters. A representation of the ancestry of an individual or family; a family tree.

pedomorphosis. The retention of infantile, fetal or embryonic characteristics into the adult.

penetrance. (1) The proportion of individuals with a given genotype (heterozygotes for a dominant gene) who express an expected trait, even if mildly. If a disease gene is not causing the disease in all its carriers, its penetrance is low [not to be mixed with variable expression]. BRCA1 mutations show both age-dependent penetrance and overall reduced penetrance, the lifetime risk for a female mutation carrier being estimated at around 70%. Breast cancer is also an example of an autosomal condition where penetrance is sex-dependent. While male mutation carriers can develop breast cancer (particularly with BRCA2 mutations), females are at much greater risk. (2) The proportion of individuals with a specific genotype who manifest that genotype at the phenotype level.

peptide bond. A bond joining two amino acids.

peptide. See amino acid and peptide bond.

peptidyl transferase. The enzymatic centre in the ribosome responsible for peptide bond formation during translation.

pericentric inversion. An inversion that involves the centromere. A chromosomal inversion that includes the centromere.

permissive condition. Environmental condition under which a conditional mutation shows the wild-type phenotype.

permissive temperature. A temperature at which temperature-sensitive mutants are normal.

PEST hypothesis. The short life-time of a protein is signaled by a region rich in the amino acids proline (P); glutamic acid (E); serine (S); or threonine (T).

petite mutation. A mutation of yeast that produces small (petite) anaerobic-like colonies.

petite. A yeast mutation producing small colonies and altered mitochondrial functions. In cytoplasmic (neutral and suppressive) petites the mutation is a deletion in mitochondrial DNA, while in segregational petites the mutation occurs in nuclear DNA.

phage. See bacteriophage.

phagemid. A type of plasmid which carries within its sequence a bacteriophage replication origin. When the host bacterium is infected with "helper" phage, the phagemid is replicated along with the phage DNA and packaged into phage capsids.

Phe. Phenylalanine (an amino acid).

phenocopy. (1) A condition which is due to environmental factors but resembles one which is genetic. (2) A phenotype that is not genetically controlled but looks like a genetically controlled phenotype. An environmentally induced phenotype that resembles the phenotype produced by a mutation.

phenotype. (1) The form taken by some character (or group of characters) in a specific individual. (2) The detectable outward manifestations of a specific genotype. (3) The observable attributes of an organism. (4) The visible or measurable (i.e., expressed) characteristics of an organism (see genotype).

phenotypic sex determination. Sex determination by non-genetic means.

phenotypic variance. Variance of phenotype due to genotypic and environmental factors combined.

phenylketonuria(pku). A human metabolic disease caused by a mutation in a gene coding for a phenylalanine processing enzyme (phenylalanine hydroxylase), which leads to accumulation of phenylalanine and mental retardation if not treated; inherited as an autosomal recessive phenotype.

pheromone (formerly called mating-type factors, sex factors or gamones). (1) Species- or mating-type-specific chemical produced by an animal to communicate with an effect on their behavior without being consciously perceived as smell. Probably the most ancient communication system in living organisms. It has been noted in bacteria (Streptomyces faecalis); protists such as ciliates (Euplotes raikovi); amoeba (Dictyostelium); algae (Fucus vesiculosus); and in fungi (for example Basidiomycetous). It is mainly used to attract opposite sex in insects (in cockroaches, moths, beetles and bees) but social insects use them for communication other than reproduction, and mammals use them for territory marking. In vertebrates, they are also used for kin recognition and mate selection. Specific examples of pheromones include a substance produced by male cockroaches that orients females in the correct mating positions and a desert locust pheromone that accelerates sexual maturation in adults of both sexes. Male-attracting pheromones are produced by the females of many species of beetles, bees, and moths. The polyphemus moth needs red oak leaves for mating as the leaves release a volatile aldehyde that stimulates the female to produce her male-attracting pheromone. (2) A chemical signal, analogous to a hormone, that passes information between individuals.

Philadelphia chromosome. A translocation between the long arms of chromosomes 9 and 22, often found in the white blood cells of patients with chronic myeloid leukemia.

phosphodiester bond. A bond between a two sugar groups and a phosphate group; such bonds form the sugar-phosphate-sugar backbone of DNA and RNA. A diester bond (between phosphoric acid and two sugar molecules) linking two nucleotides together to form the nucleotide polymers DNA and RNA.

photoreactivation. The process whereby dimerized pyrimidines (usually thymines) in DNA are restored by an enzyme (deoxyribodipyrimidine photolyase) that requires light energy.

phyletic evolution. See anagenesis.

phyletic gradualism. (1) A model of evolutionary mode characterized by slow and gradual modifications of biological structures leading to speciation. This is the opposite of punctuated equilibrium. (2) The process of gradual evolutionary change over time.

phylogenetic footprinting. The use of phylogenetic comparisons to reveal conserved functional elements.

phylogenetic tree. A diagram showing evolutionary lineages of organisms.

phylogenetics. Study of reconstructing evolutionary genealogical ties between taxa and line of descent of species or higher taxon.

phylogeny. An evolutionary tree showing the inferred relationships of descent and common ancestry of any given taxa.

piebald. A mammalian phenotype in which patches of skin are unpigmented because of lack of melanocytes; generally inherited as an autosomal dominant.

pilus (plural pili). A conjugation tube; a hollow hair-like appendage of a donor Escherichia coli cell that acts as a bridge for transmission of donor DNA to the recipient cell during conjugation.

plant breeding. The application of genetic analysis to development of plant lines better suited for human purposes.

plaque. A clear area on a bacterial lawn, left by lysis of the bacteria through progressive infections by a phage and its descendants. Clear area on a bacterial lawn caused by cell lysis due to viral attack.

plasmid. (1) A circular piece of DNA present in bacteria or isolated from bacteria. *Escherichia coli*, the usual bacteria in molecular genetics experiments, has a large circular genome, but it will also replicate smaller circular DNAs as long as they have an "origin of replication". Plasmids may also have other DNA inserted by the investigator. A bacterium carrying a plasmid and replicating a million-fold will produce a million identical copies of that plasmid. Common plasmids are pBR322, pGEM, pUC18. (2) A transferable extrachromosomal genetic element found in some bacteria. They are 1 to 200 kb long, double-stranded, circular DNA molecules. They replicate independently of the bacterial chromosomes and usually confer an advantage to the bacteria (such as antibiotic or heavy metal resistance). Plasmids are popular vectors in recombinant DNA technology. They can carry up to 10 kb foreign DNA. (3) Autonomously replicating extrachromosomal DNA molecule. An autonomous self-replicating genetic particle usually of circular double-stranded DNA.

plastid. A chloroplast prior to the development of chlorophyll.

plastid. Chloroplast and related organelles in the cells of eukaryotic algae and green plants. Together with mitochondria they are responsible for energy production.

plate. (1) A flat dish used to culture microbes. (2) To spread cells over the surface of solid medium in a plate.

platyrrhini. New world monkeys found in South America with widely spaced nostrils. They last had a common ancestor with the old world monkeys (Catarrhini) about 55 million years ago.

pleiotropic mutation. A mutation that has effects on several different characters.

pleiotropy. (1) More than one effect of a gene on the phenotype. The effects may occur simultaneously or sequentially. An example is the determination of the color pattern and the shape of the eyes by a single allele in Siamese cats. Another example is the DNA repair enzymes which have several other functions (transcription, cell cycle regulation, regulation of gene rearrangements). (2) The phenomenon whereby a single mutation affects several apparently unrelated aspects of the phenotype.

point mutation. (1) A mutation that can be mapped to one specific site within a locus. A small mutation that consists of the replacement (transition or transversion); addition; or deletion (frameshift) of one or a few bases. (2) A single nucleotide change in the DNA sequence. Even if it is in the coding region of a gene, it may or may not change the amino acid sequence. The rates of point mutation for MHC genes are not unusually high. The extensive polymorphism results from their accumulation over many millions of years of transspecies evolution.

Poisson distribution. A mathematical expression giving the probability of observing various numbers of a particular event in a sample when the mean probability of that event on any one trial is very small.

poky mutation. A mutation of Neurospora that produces a petite-like phenotype.

polar body. A small cell (which eventually disintegrates) that is the by-product of meiosis in female animals.

One functional ovum and potentially three polar bodies result from meiosis of each primary oocyte.

polar gene conversion. A gradient of conversion frequency along the length of a gene.

polar granules. Cytoplasmic granules localized at the posterior end of a Drosophila oocyte and early embryo. These granules are associated with determinants of the germ-line and posterior structures.

polar mutation. A mutation that affects the transcription or translation of part of the gene or operon downstream of the mutant site. For example, nonsense mutations, frameshift mutations, and insertion sequence(IS)-induced mutations.

polarity gene. A mitochondrial gene with alleles that are preferentially found in daughter mitochondria after a recombinational event between mitochondria.

polarity. Meaning directionality and referring either to an effect seen in only one direction from a point of origin or to the fact that linear entities (such as a single strand of DNA or RNA or a protein) have ends (5' and 3'; N and C) that differ from each other.

pollen (tube) competition. The plant equivalent of sperm competition (see also cryptic female selection). A type of sexual selection.

pollen grain. (1) The male gametophyte in higher plants. (2) The microspores of seed plants. It germinates to form the male gametophyte. The male gametophyte contains three haploid nuclei. One of these fertilizes the ovum, a second fuses with the two polar nuclei to form the triploid endosperm, and the third degenerates once double fertilization occurs.

poly (A) tail. A sequence of 20 - 200 adenylic acid residues which is added to the 3' end of most eukaryotic mRNAs. It increases their stability by making them resistant to nuclease digestion.

poly(A) tail. (1) A string of adenine nucleotides added to the 3' end of eukaryotic mRNA after transcription. (2) After an mRNA is transcribed from a gene, the cell adds a stretch of A residues (typically 50-200) to its 3' end. It is thought that the presence of this "polyA tail" increases the stability of the mRNA (possibly by protecting it from nucleases). Note that not all mRNAs have a polyA tail; the histone mRNAs in particular do not.

polyacrylamide. A material used to make gels for separation of mixtures of macromolecules by electrophoresis.

polycistronic mRNA. An mRNA that codes for more than one protein.

polycistronic. Referring to prokaryotic messenger RNAs that contain several cistrons within the same mRNA transcript.

poly-dA/poly-dT technique. A method of inserting DNA into a vector by adding poly-dA to the linearised vector and poly-dT to the DNA of interest. Also feasible with poly-dG and poly-dC.

polydactyly. More than five fingers and/or toes. Inherited as an autosomal dominant phenotype.

polygene. See multiple-factor hypothesis.

polygenic inheritance. See quantitative inheritance.

polygenic. Traits controlled by two or more genetic loci. They are usually influenced by environment as well (multifactorial).

polymerase chain reaction (PCR). (1) A powerful method for amplifying specific DNA segments which exploits certain features of DNA replication. For instance replication requires a primer and specificity is determined by the sequence and size of the primer. The method amplifies specific DNA segments by cycles of template denaturation; primer addition; primer annealing and replication using thermostable DNA polymerase. The degree of amplification achieved is set at a theoretical maximum of 2^N, where N is the number of cycles, e.g., 20 cycles gives a theoretical 1048576 fold amplification. (2) A technique for replicating a specific piece of DNA *in-vitro* , even in the presence of excess non-specific DNA. Primers are added (which initiate the copying of each strand) along with nucleotides and Taq polymerase. By cycling the temperature, the target DNA is repetitively denatured and copied. A single copy of the target DNA, even if mixed in with other undesirable DNA, can be amplified to obtain billions of replicates. PCR can be used to amplify RNA sequences if they are first converted to DNA via reverse transcriptase. This two-phase procedure is known as 'RT-PCR'.

Polymerase Chain Reaction (PCR) is the basis for a number of extremely important methods in molecular biology. It can be used to detect and measure vanishingly small amounts of DNA and to create customized pieces of DNA. It has been applied to clinical diagnosis and therapy, to forensics and to vast numbers of research applications. It would be difficult to overstate the importance of PCR to science.

polymerase. An enzyme which links individual nucleotides together into a long strand, using another strand as a template. There are two general types of polymerase — DNA polymerases (which synthesize DNA) and RNA polymerase (which makes RNA). Within these two classes, there are numerous sub-types of polymerase, depending on what type of nucleic acid can function as template and what type of nucleic acid is formed. A DNA-dependant DNA polymerase will copy one DNA strand starting from a primer, and the product will be the complementary DNA strand. A DNA-dependant RNA polymerase will use DNA as a template to synthesize an RNA strand.

polymerize. To form a complex compound (polymer) by linking together many smaller elements (residues).

polymorphism. (1) Presence of discreetly different forms of a gene or a character. It is defined as a Mendelian trait that exists in the population in at least two phenotypes, neither of which occurs at a frequency of less than 1%. Polymorphism at a genetic locus is due to either balanced polymorphism (heterozygous advantage, frequency-dependent selection) or unequilibrium states (temporary polymorphism) as occurs during frequency-dependent selection and genetic drift (alleles becoming fixed or extinct). (2) The occurrence in a population (or among populations) of several phenotypic forms associated with alleles of one gene or homologs of one chromosome. See genetic polymorphism.

polynucleotide phosphorylase. An enzyme that can polymerize nucleotide diphosphates without the need for a primer. The function of this enzyme in vivo is probably in its reverse role as an RNA exonuclease.

polypeptide. A chain of linked amino acids; a protein.

polyploid. (1) A cell or an organism having three or more chromosome sets. (2) The situation in which the organism has more than two (2n) sets of chromosomes. It could be 3n, 4n or more. A common situation in earthworms and plants. About half of angiosperms are polyploid. It arises as a result of meiotic irregularities and gives rise to sterile progeny which can still reproduce asexually. The original South American potato is a tetraploid (4n). Many of the common food plants are polyploid as this results in larger flowers and fruits (as well as larger cells, thicker and fleshier leaves). The wheat now grown for bread T. aestivum is hexaploid (6n = 42 chromosomes). Polyploidy is a common mechanism for sympatric speciation which played a role in angiosperm evolution. Link to a mini-essay on polyploidy.

polyribosome. See polysome.

polysaccharide. A biological polymer composed of sugar subunits for example, starch or cellulose.

polysome. The configuration of several ribosomes simultaneously translating the same mRNA. Shortened form of the term polyribosome.

polytene chromosome. A giant chromosome produced by an endomitotic process in which, following synapsis of the two homologues, multiple rounds of replication produce chromatids that remain synapsed together in a haploid number of chromosomes. Large chromosome consisting of many chromatids formed by rounds of endomitosis following synapsis of the two homologues.

population biology. The study of the patterns in which organisms are related in time and space. It is a combination of disciplines such as population genetics, ecology, taxonomy, ethology and others.

population genetics. The branch of genetics that deals with frequencies of alleles and genotypes in breeding populations. It also deals with selective influences on the genetic composition of the population

population. A group of organisms of the same species relatively isolated from other groups of the same species. See deme.

position effect. (1) A difference in phenotype that is dependent on the position of a gene or a group of genes, often caused by heterochromatin nearby. Thus, the change in a gene's location may cause a change in its expression (a problem that has to be overcome in gene therapy). (2) Used to describe a situation in which the phenotype expressed by a gene is altered by changes in the position of the gene within the genome. For instance genes translocated to regions of heterochromatin are often not expressed.

position-effect variegation. Variegation caused by the inactivation of a gene in some cells through its abnormal juxtaposition with heterochromatin.

positive assortative mating. (1) A situation in which like phenotypes mate more commonly than expected by chance. (2) A type of nonrandom mating in which individuals of similar phenotype mate more often than predicted under random mating conditions.

positive control. Regulation mediated by a protein that is required for the activation of a transcription unit.

positive interference. When the occurrence of one crossover reduces the probability that a second will occur in the same region .

postreplicative repair. A DNA repair process initiated when DNA polymerase bypasses a damaged area. enzymes in the rec system are used.

post-transcriptional modification. Changes in eukaryotic mRNA, tRNA or other RNAs made after transcription has been completed. The changes to mRNA include addition of a 5'cap and 3' polyA tail and removal of introns, and to tRNA include modification of bases and removal of introns.

post-transcriptional regulation. Any process occurring after transcription which affects the amount of protein a gene produces. Includes RNA processing efficiency, RNA stability, translation efficiency, protein stability. For example, the rapid degradation of an mRNA will reduce the amount of protein arising from it.

Increasing the rate at which an mRNA is translated will increase the amount of protein product.

post-translational modifications. Cleavage of amino terminal peptide, hydroxylation and oxidation of amino acids in the polypeptide chain for cross-linking, covalent modifications by acetylation, phosphorylation and glycosylation.

post-translational processing. The reactions which alter a protein's covalent structure, such as phosphorylation, glycosylation or proteolytic cleavage.

post-translational regulation. Any process which affects the amount of protein produced from a gene, and which occurs AFTER translation in the grand scheme of genetic expression. Actually, this is often just a buzz-word for regulation of the stability of the protein. The more stable a protein is, the more it will accumulate.

PRE (Progesterone Response Element). A binding site in a promoter to which the activated progesterone receptor can bind. The progesterone receptor is essentially a transcription factor which is activated only in the presence of progesterone . The activated receptor will bind to a PRE, and transcription of the adjacent gene will be altered. See also "Response element".

pre-Cambrian Eon. The whole of geological time before the Cambrian period (< 540 Mya).

preemptor stem. A configuration of leader transcript mRNA that allows transcription in attenuator-controlled amino acid operons.

pre-mRNA (precursor mRNA). (1) The primary transcript and intermediates in RNA processing that yield functional (mature) mRNA. (2) The first (primary) transcript from a protein coding gene is often called a pre-mRNA and contains both introns and exons. Pre-mRNA requires splicing (removal) of introns to produce the final mRNA molecule containing only exons.

Pribnow box. Relatively invariant sequence of six nucleotides (concensus TATAAT) in prokaryotic promoters centred at the position minus 10 (10 base pairs upstream from the transcription start site).

primary oocyte. A cell that undergoes meiosis in female animals.

primary spermatocyte. A cell that undergoes the first meiotic division in male animals. See secondary spermatocyte.

primary structure. The sequence of polymerized amino acids in a protein.

primary transcript. (1) The product of eukaryotic transcription before post-transcriptional modifications take place. (2) When a gene is transcribed in the nucleus, the initial product is the primary transcript, an RNA containing copies of all exons and introns. This primary transcript is then processed by the cell to remove the introns, to cleave off unwanted 3' sequence, and to polyadenylate the 5' end. The mature message thus formed is then exported to the cytoplasm for translation.

primase. (1) An enzyme that creates an RNA primer for initiation of DNA replication. (2) An enzyme that makes the RNA primer required by DNA polymerase in DNA replication. See also primosome.

primates. One of the Mammalian Orders which includes Lemurs (suborder Prosimii), old world and new world Monkeys, Apes and Humans. The suborder Anthropoidea covers all Primates except Prosimii. Prosimians and Anthropoids diverged from each other 65 Mya; Apes and Old World Monkeys 35 Mya. The separation of the two groups of Anthropoids (Platyrrhini and Catarrhini) occurred about 55 Mya.

primer extension. This is a method used to figure out how far upstream from a fixed site the start of an mRNA is. For example, perhaps you have isolated a cDNA clone, but you don't think that the clone has all of the 5' untranslated region. To find out how much is missing, you would first sequence the part you have, and figure out which strand is coding strand (usually the coding strand will have a large open reading frame). Next, you ask the DNA Synthesis Facility to make an oligonucleotide complementary to the 5'-most region of the coding strand (and thus complementary to the mRNA). This "primer" is hybridized to mRNA (say, a mixture of mRNA containing the one in which you are interested), and reverse transcriptase is added to copy the mRNA from the primer out to the 5' end. The size of the resulting DNA fragment shows how far away from the 5' end your primer is.

primer. (1) A short nucleic acid molecule which, when annealed to a complementary template strand, provides a 3' terminus suitable for copying by a DNA polymerase. (2) A short sequence (of RNA or DNA) from which DNA replication can initiate. May be either a synthetic DNA or RNA or a length of RNA synthesized in vivo by primase. (3) A small oligonucleotide (anywhere from 6 to 50 nt long) used to prime DNA synthesis. The DNA polymerases are only able to extend a pre-existing strand along a template; they are not able to take a naked single strand and produce a complementary copy of it *de-novo*. A primer which sticks to the template is therefore used to initiate the replication. Primers are necessary for DNA sequencing and PCR.

primosome. (1) A complex of two proteins, a primase and a helicase, that initiates RNA primers on the lagging DNA strand during DNA replication. (2) The mobile complex of helicase and primase that is involved in DNA replication.

prion. (1) An infectious agent which does not have any nucleic acid (but just protein). Responsible for scrapie in sheep, kuru and Creutzfeldt-Jakob disease in humans. (2) Protein infectious agent associated with several neurological diseases (scrapie; kuru; Creutzfeld-Jakob syndrome; Alzheimer's disease). Each disease has a different prion.

Pro. Proline (an amino acid).

probability theory. The conceptual framework concerned with quantification of probabilities. See probability.

probability. The expectation of the occurrence of a particular event.

proband. See propositus.

probe. (1) A fragment of DNA or RNA which is labeled in some way (often incorporating 32P or 35S), and which is used to hybridize with the nucleic acid in which you are interested. For example, if you want to quantitate the levels of alpha subunit mRNA in a preparation of pituitary RNA, you might make a radiolabeled RNA *in-vitro* which is complementary to the mRNA, and then use it to probe a Northern blot of the pit RNA. A probe can be radiolabeled, or tagged with another functional group such as biotin. A probe can be cloned DNA, or might be a synthetic DNA strand. As an example of the latter, perhaps you have isolated a protein for which you wish to obtain a cDNA or genomic clone. You might (pay to) microsequence a portion of the protein, deduce the nucleic acid sequence, (pay to) synthesize an oligonucleotide carrying that sequence, radiolabel it and use it as a probe to screen a cDNA library or genomic library. A better way is to call up someone who already has the clone. (2) Defined nucleic acid (DNA or RNA) that can be used to identify, usually through autoradiography, specific DNA or RNA molecules bearing the complementary sequence. A labeled (radioactive; antigen; enzyme; fluorescent) nucleic acid complementary to the sequence being searched for in a restriction digest, genome library, northern blot or in situ hybridization.

processed pseudogene. A pseudogene, derived from a retrotranscript of mRNA of any expressed gene and inserted back into the genome. A processed pseudogene is intronless, usually flanked by the repeat sequence (GC/AGCTCTCC), and rich in multiple genetic lesions including substitution, deletion and/or insertion events that modify the reading frames. Because of the lack of its original promoter and the genetic lesions it has accumulated, a processed pseudogene is not normally expressed.

processing. The reactions occurring in the nucleus which convert the primary RNA transcript to a mature mRNA. Processing reactions include capping, splicing and polyadenylation. The term can also refer to the processing of the protein product, including proteolytic cleavages, glycosylation, etc.

processivity. The ability of an enzyme to repetitively continue its catalytic function without dissociating from its substrate.

product of meiosis. One of the (usually four) cells formed by the two meiotic divisions.

product rule. The probability of two independent events occurring simultaneously is the product of the individual probabilities. The rule stating that the probability of the occurrence of independent events is the product of their separate probabilities.

proflavin. A mutagen that tends to produce frameshift mutations.

progeny testing. Breeding of offspring to determine their genotypes and that of their parents.

prokaryote. An organism lacking a true nucleus, such as a bacterium or a blue-green alga.

prokaryotic cell. (1) A cell having no nuclear membrane and hence no separate nucleus. (2) The cell type in which the DNA is not enclosed in a nucleus. Consists of Eubacteria and Archaebacteria. Always unicellular. When the cell has a proper nucleus (eu-karyon), it is eukaryotic.

promoter. (1) A regulatory region a short distance upstream from the 5' end of a transcription start site that acts as the binding site for RNA polymerase. A region of DNA to which RNA polymerase binds in order to initiate transcription. (2) Initial binding site for RNA polymerase in the process of gene expression. First transcription factors bind to the promoter which is located 5' to the transcription initiation site in a gene. General and tissue/cell-specific promoters stimulate the expression of a gene under the control of enhancers. (3) The first few hundred nucleotides of DNA "upstream" (on the 5' side) of a gene, which control the transcription of that gene. The promoter is part of the 5' flanking DNA, i.e. it is not transcribed into RNA, but without the promoter, the gene is not functional. Note that the definition is a bit hazy as far as the size of the region encompassed, but the "promoter" of a gene starts with the nucleotide immediately upstream from the cap site, and includes binding sites for one or more transcription factors which can not work if moved farther away from the gene.

promoter-proximal element. Any regulatory sequence in eukaryotic DNA that is located within 200 bp of a **promoter** and binds a specific protein to modulate transcription of the associated gene.

proofread. Literally to read for the purpose of detecting errors for later correction. DNA polymerase has 3' to

5' exonuclease activity which is used during polymerization to remove nucleotides it has recently added if they are incorrectly paired, and is a correcting ability to remove errors in replication.

proofreading. In DNA synthesis, the ability of DNA polymerase to recognize mismatched bases. DNA polymerase corrects mistakes with its exonuclease activity. RNA editing is also possible at the mRNA level in some simple organisms.

prophage. (1) A phage chromosome inserted as part of the linear structure of the DNA chromosome of a bacterium. A temperate phage integrated into the host chromosome. (2) The phage DNA inserted into a bacterial chromosome.

prophase. The early stage of nuclear division during which chromosomes condense and become visible. The initial stage of mitosis or meiosis in which chromosomes become visible and the spindle apparatus forms.

proplastid. Mutant plastids that do not grow and develop into chloroplasts.

propositus. The person through whom a pedigree was discovered.

prosimians (Prosimii). The suborder of Primates including Lemurs, Lorises and Tarsiers.

protein clock hypothesis. The idea that amino acid replacements occur at a constant rate in a given protein family (ribosomal proteins, cytochromes, etc.) and the degree of divergence between two species can be used to estimate the time elapsed since their divergence.

protochordata. A division of chordata phylum including subphylum hemichordata, urochordata (tunicates) and cephalochordata (lancelets). Immediate ancestors of vertebrates.

protoctista. The modern name for the Kingdom Protista. Probably the first double-chromosomed beings. Well-known members are: seaweeds (algae), amoebas, ciliates, mastigotes, water molds, slime molds and slime nets.

proto-oncogene. (1) A gene present in a normal cell which carries out a normal cellular function, but which can become an oncogene under certain circumstances. The prefix "c-" indicates a cellular gene, and is generally used for proto-oncogenes (examples: *c-myb* , *c-myc* , *c-fos* , *c-jun* , etc). (2) The non-activated form of a cellular oncogene in an untransformed cell. A gene that, when mutated or otherwise affected, becomes an oncogene.

protoplast. A plant cell whose wall has been removed.

prototroph. A strain of organisms that will proliferate on minimal medium (compare auxotroph).

protozoa (meaning "first animals"). A collective name for several phyla of unicellular and eukaryotic, the most animal-like, mainly parasitic organisms belonging to the Kingdom Protista (modern name: Protoctista meaning "first beings"). This Kingdom includes Protozoa and other not necessarily single-celled but eukaryotic organisms (algae and others formerly known as fungi) except fungi. They reproduce by fission or conjugation, and move by cilia, flagella or pseudopodia. The taxonomic Kingdom Protista is thought to be the ancestor of the other eukaryotic kingdoms Fungi, Plants and Animals. Familiar examples of protozoa are flagellates (incl. Chlamydomonas), Amoeba, Plasmodium, Ciliates (incl. Paramecium).

provirus. A virus chromosome integrated into the DNA of the host cell.

pseudoallele. Allele that is functionally but not structurally allelic, that is wild-type recombinants can be recovered by intragenic recombination from heterozygotes containing two different pseudoalleles.

pseudoalleles. Genes that behave like alleles but can be separated by crossing over. The eye color genes on the X chromosome of Drosophila are for example closely adjacent but separable loci and not alleles of a single gene.

pseudoautosomal gene. A gene that occurs on both sex-determining heteromorphic chromosomes.

pseudoautosomal inheritance. The X and Y chromosomes share a common ancestor. There is a part of X chromosome which has its homologous counterpart on the Y chromosome. The pattern of inheritance for a gene located on both the X and Y chromosomes may appear to be autosomal.

pseudodominance. The sudden appearance of a recessive phenotype in a pedigree, due to deletion of a masking dominant gene. The phenomenon in which a recessive allele shows itself in the phenotype when only one copy of the allele is present, as in hemizygous alleles or in deletion heterozygotes.

pseudogene. (1) An inactive gene derived from an ancestral active gene. (2) A gene which has acquired a nonsense mutation and lost its transcription ability.

pseudomonas syringae. The genetically engineered strain of this bacteria lack a cell-surface protein that helps ice crystals to form. Spraying these bacteria on crops may prevent freezing-related damages.

puff. See chromosome puff.

pulse field gel electrophoresis (PFGE). A form of gel electrophoresis that allows extremely long DNA molecules to be separated from one another.

pulse-chase experiment. An experiment in which cells are grown in radioactive medium for a brief period (the pulse) and then transferred to nonradioactive medium for a longer period (the chase).

pulsed field gel electrophoresis (PFGE). (1) A gel technique which allows size-separation of very large fragments of DNA, in the range of hundreds of kb to thousands of kb. As in other gel electrophoresis techniques, populations of molecules migrate through the gel at a speed related to their size, producing discrete bands. In normal electrophoresis, DNA fragments greater than a certain size limit all migrate at the same rate through the gel. In PFGE, the electrophoretic voltage is applied alternately along two perpendicular axes, which forces even the larger DNA fragments to separate by size. (2) An electrophoretic technique in which the gel is subjected to electrical fields alternating between different angles, allowing very large DNA fragments to snake through the gel, and hence permitting efficient separation of mixtures of such large fragments.

punctuated equilibrium. (1) Put forward by Niles Eldredge and Stephen Jay Gould in 1972 as a counter theory to Darwin's gradualism in speciation (see phyletic gradualism). It suggests that new species may have arisen rapidly over a few thousand years and then remained unchanged (stasis) for many millions of years. Punctuated equilibrium postulates that change occurred in only a small part of the population (rather than the whole population is evolving gradually). The most plausible explanation for a sudden and drastic change would be mutations in regulatory sequences that affect a whole operon (see Gould SJ & N Eldredge. Punctuated equilibria: the tempo and mode of evolution reconsidered. Paleobiology 1977; 3:115-51). (2) The evolutionary process involving long periods without change (stasis) punctuated by short periods of rapid speciation.

Punnett square. A diagrammatic representation of a particular cross used to predict the progeny of the cross. A grid used as a graphic representation of the progeny zygotes resulting from different gamete fusions in a specific cross.

pure-breeding line or strain. A group of identical individuals that always produce offspring of the same phenotype when intercrossed.

purines. A type of nitrogen base; the purine bases in DNA and RNA are adenine and guanine.

pyrimidine. Nitrogenous bases of which thymine is found in DNA; uridine in RNA; and cytosine in both.

Q

quantitative character. A character displaying a continuous phenotypic range rather than discrete classes (characters taking any value within a limit; characters measured rather than counted such as metabolic activity, height, length, width, body fat content, growth rate, milk production). The genetic variation underlying a continuous character distribution may be the result of segregation at a single genetic locus or more frequently, at numerous interacting loci which produce cumulative effect on the phenotype. A gene affecting a quantitative character is a quantitative trait locus, or QTL.

quantitative genetics. The statistical study of the genetics of quantitative characters (biometrical genetics) as opposed to Mendelian (discrete) characters. Quantitative genetic characters are those that do not assort in a simple way in crosses. Examples are physiological activity, reproductive rate, behavior, size, and height. A major task of quantitative genetics is to determine the ways in which genes interact with the environment to contribute to the formation of a given quantitative trait distribution (and the estimation of genetic and environmental variance). Link to a lecture on quantitative genetics.

quantitative inheritance. The mechanism of genetic control of traits showing continuous variation.

quantitative variation. Variation measured on a continuum (e.g., height in human beings) rather than in discrete units or categories. See continuous variation. The existence of a range of phenotypes for a specific character, differing by degree rather than by distinct qualitative differences.

quasidominance. Direct transmission, generation to generation, of a recessive trait giving the impression of dominance. It happens if the recessive gene is frequent or inbreeding is intense.

quasispecies. The whole population of phylogenetically related (virus) variants observed within a single (infected) individual. Viruses with high mutation rates such as human immunodeficiency virus (HIV) and hepatitis C virus (HCV) occur like this. As a comparison, HIV variation within a single infected individual can be as great as the variation of influenza throughout the worldwide-infected population in a flu season.

quaternary structure. The association of polypeptide subunits to form the final structure of a protein.

R

R factor. See R plasmid.

R plasmid. A plasmid containing one or several transposons with resistance genes. Plasmids carrying genes controlling resistance to various drugs.

race. Described in population genetics as a geographic subdivision of a species distinguished from others by the allele frequencies of a number of genes. A beautiful discussion that there are no genetically defined races within Homo sapiens can be found in Cavalli-Sforza's book *Genes, Peoples, and Languages* (2000).

RAM mutation. Ribosomal ambiguity mutation that allows incorrect tRNAs to be incorporated into the translation process.

random genetic drift. Changes in allelic frequency due to sampling error. Changes in allele frequency that result because the genes appearing in offspring are not a perfectly representative sampling of the parental genes (e.g. in small populations). See also founder effect.

random mating. Mating between individuals where the choice of partner is not influenced by the genotypes (with respect to specific genes under study). The mating of individuals in a population such that the union of individuals with the trait under study occurs according to the product rule of probability.

random mating. Mating without any preference for mates. One of the assumptions of Hardy-Weinberg equilibrium (nonrandom mating may be due to assortative or disassortative mating).

random primed synthesis. If you have a DNA clone and you want to produce radioactive copies of it, one way is to denature it (separate the strands), then hybridize to that template a mixture of all possible 6-mer oligonucleotides. Those oligos will act as primers for the synthesis of labeled strands by DNA polymerase (in the presence of radiolabeled precursors).

random strand analysis. Mapping studies in organisms that do not keep together all the products of meiosis.

reading frame. (1) The succession of codons determined by reading nucleotides in groups of three from a specific initiation codon. (2) When mRNA is translated by the cell, the nucleotides are read three at a time. By starting at different positions, the groupings of three that are produced can be entirely different. The following example shows a DNA sequence and the three reading frames in which it could be read. Not only is an entirely different amino acid sequence specified by the different reading frames, but two of the three frames have stop codons, and thus are not open reading frames (asterisks indicate a stop codon).

readthrough. Transcription or translation beyond the normal termination signals in DNA or mRNA respectively.

realized heritability. Heritability measured by a response to selection. The ratio of the single-generation progress of selection to the selection differential of the parents.

reannealing. Spontaneous realignment of two single DNA strands to re-form a DNA double helix.

rec system. Several loci controlling genes (recA; recB; recC and others) involved in postreplicative DNA repair.

receptor element. A controlling element in maize that can insert into a gene (making it mutant) and that can also excise (thus making the mutation unstable); both of these functions are nonautonomous, being under the influence of a regulator element.

recessive allele. An allele whose phenotypic effect is not expressed in a heterozygote.

recessive phenotype. The phenotype of a homozygote for the recessive allele; the parental phenotype that is not expressed in a heterozygote.

recessive. (1) A trait that is not expressed in heterozygotes (i.e., that can only be expressed in the homozygous state). (2) An allele that is not expressed in the heterozygous condition. Also the phenotype of the homozygote of a recessive allele.

reciprocal altruism. An apparently altruistic behavior performed with the understanding that the recipient will reciprocate at some future date.

reciprocal cross. A cross, with the phenotype of each sex reversed as compared with the original cross, to test the role of parental sex on inheritance pattern. A pair of crosses of the type genotype A(female) X genotype B(male) and genotype B(female) X genotype A(male).

reciprocal translocation. A chromosomal configuration in which (usually) the ends of two non-homologous chromosomes have become exchanged. A translocation in which part of one chromosome is exchanged with a part of a separate non-homologous chromosome.

recombinant DNA technology. Techniques of gene cloning. Recombinant DNA refers to the molecule formed by joining a DNA of interest to vector DNA. See gene cloning.

recombinant DNA. A novel DNA sequence formed by the joining, usually **in vitro**, of two non-homologous DNA molecules.

recombinant frequency (RF). The proportion (or percentage) of recombinant cells or individuals.

recombinant. In genetic mapping studies an offspring having a non-parental allele combination. An individual or cell with a genotype produced by recombination.

recombinase. A group of enzymes that catalyze the joining of two DNA molecules after recognizing the recombination sites. See also integrase and transposase.

recombination (crossing-over). (1) The exchange (reshuffling) of genetic material between a homologous pair of chromosomes during meiosis (see also somatic recombination; sister chromatid exchange). (2) The non-parental arrangement of alleles in progeny that can result from either independent assortment or crossing over. (3) In general, any process in a diploid or partially diploid cell that generates new gene or chromosomal combinations not found in that cell or in its progenitors. (4) At meiosis, the process that generates a haploid product of meiosis whose genotype is different from either of the two haploid genotypes that constituted the original meiotic diploid.

recombinational repair. The repair of a DNA lesion through a process, similar to recombination, that uses recombination enzymes.

recon. A term coined by Seymour Benzer for the smallest recombinable unit within a cistron. A region of a gene within which there can be no crossing-over; now known to be a single nucleotide pair.

reduction division. A nuclear division that produces daughter nuclei each having one-half as many centromeres as the parental nucleus. The first meiotic division reduces the number of chromosomes and centromeres to half that of the original cell.

redundant DNA. See repetitive DNA.

regression coefficient. The slope of the straight line that most closely relates two correlated variables.

regression. A term coined by Galton for the tendency of the quantitative traits of offspring to be closer to the population mean than are their parents' traits. It arises from a combination of factors - dominance, gene interactions, and environmental influences on traits.

regulator element. See receptor element.

regulator gene. A gene primarily involved in controlling another (structural) gene. Genes that are involved in turning on or off the transcription of structural genes.

relative Darwinian fitness. See fitness.

relaxed mutant. A mutant that does not exhibit the stringent response under amino acid starvation.

release factors (RFI and RF2). Proteins in prokaryotes responsible for termination of translation and release of the newly synthesized polypeptide when a nonsense codon appears in the A site of the ribosome. Replaced by eRF in eukaryotes.

Renner complex. Specific gametic chromosome combination in Oenothera the evening primrose.

repetitive DNA. (1) A surprising portion of any genome consists not of genes or structural elements, but of frequently repeated simple sequences. These may be short repeats just a few nt long, like CACACA etc. They can also range up to a few hundred nt long. Examples of the latter include Alu repeats, LINEs,

SINEs. The function of these elements is often unknown. In shorter repeats like di- and tri-nucleotide repeats, the number of repeating units can occasionally change during evolution and descent. They are thus useful markers for familial relationships and have been used in paternity testing, forensic science and in the identification of human remains. (2) DNA made up of copies of the same or nearly the same nucleotide sequence. DNA sequences that are present in many copies per chromosome set. Repetitive DNA sequences may be closely linked (e.g., satellite DNA or VNTR loci) or dispersed throughout the genome or parts of the genome (e.g., alu family). (3) Non-coding DNA which consists of nucleotide sequences repeatedly occurring in chromosomal DNA. They do not normally have any function but those capping the chromosomes prevent the loss of genetic information after each replication (as this would cause a 3' overhang). In human genome, at least 20% of the DNA consists of repetitive sequences.

replica plating. A technique to rapidly transfer microorganism colonies from a master plate, in an exact spatial pattern, to a number of further plates.

replication fork. The point at which the two strands of DNA are separated to allow replication of each strand.

replication. DNA synthesis. The process of copying.

replicative transposition. The transposition of a transposable element to a new location without it being lost from the original location.

replicon. (1) A chromosomal region whose replication is controlled by a single adjacent DNA replication initiation site. A genetic unit of replication including a length of DNA and its site for initiation of replication. (2) A unit of genetic material which behaves autonomously during replication of DNA. In bacteria, a whole chromosome is a replicon. In eukaryotes, chromosomes are divided into hundreds of replicons. Each replicon contains a segment beginning with a binding site for RNA polymerase.

replisome. The DNA-replicating structure at the replication fork consisting of two DNA polymerase III enzymes and a primosome (primase and DNA helicase).

reporter gene. A gene whose phenotypic expression is easy to monitor; used to study promoter activity in different tissues or developmental stages. Recombinant DNA constructs are made in which the reporter gene is attached to a promoter region of particular interest and the construct transfected into a cell or organism.

repressible operon. Synthesis of a coordinated group of enzymes, involved in a single synthetic (anabolic) pathway, is repressible if excess quantities of (usually) the end product of the pathway leads to cessation of transcription of the genes encoding the enzymes of the pathway.

repression. The condition of an inactive gene. Lack of activity (transcription) due to presence of repressor.

repressor. The protein product of a regulator gene that acts to control transcription of inducible and repressible operons. A molecule that binds to the operator and prevents transcription of an operon. Generally any molecule that can reversibly inactivate a gene.

reproductive isolating mechanism. Any environmental; behavioural; mechanical; and physiological barrier that prevents two individuals of different populations producing viable progeny.

reproductive success. The relative production of offspring by a particular genotype.

repulsion (trans-arrangement). The condition in which a double heterozygote has received a mutant and a wild-type allele from each parent, e.g., *a* + / + *b* (see also coupling).

repulsion. Allelic arrangement of two linked heterozygous loci, in which each homologous chromosome has one mutant (a or b) and one wild-type (A or B) allele (i.e. Ab/aB). Two linked heterozygous gene pairs in the arrangement, Ab/aB.

residue. A compound such as an amino acid or a nucleotide when it is part of a larger molecule.

resistance transfer factor. See R plasmid.

resolving power. The ability of an experimental technique to distinguish between two genetic conditions (typically discussed when one condition is rare (e.g. recombination between very closely linked loci) and of particular interest) .

response element. By definition, a "response element" is a portion of a gene which must be present in order for that gene to respond to some hormone or other stimulus. Response elements are binding sites for transcription factors. Certain transcription factors are activated by stimuli such as hormones or heat shock. A gene may respond to the presence of that hormone because the gene has in its promoter region a binding site for hormone-activated transcription factor. Example, the glucocorticoid response element (GRE).

restricted transduction. See specialized transduction.

restriction digest. The result of the action of a restriction endonuclease on a DNA sample.

restriction endonuclease. Endonuclease that recognizes a particular short DNA sequence which they cleave. They help to protect cells from viral infection and are used in work with DNA.

restriction enzyme. (1) A class of enzymes ("restriction endonucleases") generally isolated from bacteria, which are able to recognize and cut specific sequences ("restriction sites") in DNA. The recognition site is thus said to be palindromic, which is typical of restriction sites. Every copy of a plasmid is identical in sequence, so if BamHI cuts a particular circular plasmid at three sites producing three "restriction fragments", then a million copies of that plasmid will produce those same restriction fragments a million times over. There are more than six hundred known restriction enzymes.

Bacteria produce restriction enzymes for protection against invasion by foreign DNA such as phages. The bacteria's own DNA is modified in such a way as to prevent it from being clipped.

(2) An endonuclease that will recognize a specific target nucleotide sequence in DNA and break the DNA chain at the target; a variety of these enzymes are known, and they are extensively used in genetic engineering.

restriction fragment length polymorphism (RFLP). Variation in DNA fragment banding patterns of electrophoresed restriction digests of DNA from different individuals of a species. Often due to the presence of a restriction enzyme cleavage site at one place in the genome in one individual and the absence of that specific site in another individual. See also variable-number-of-tandem-repeats (VNTR) locus.

restriction fragment length polymorphism. See "RFLP".

restriction fragment. The piece of DNA released after restriction digestion of plasmids or genomic DNA. See "Restriction enzyme". One can digest a plasmid and isolate one particular restriction fragment (actually a set of identical fragments). The term also describes the fragments detected on a genomic blot which carry the gene of interest.

restriction map. (1) A "cartoon" depiction of the locations within a stretch of known DNA where restriction enzymes will cut. The map usually indicates the approximate length of the entire piece (scale on the bottom), as well as the position within the piece at which designated enzymes will cut. This map happens to be of a plasmid, and the two ends are joined together with about 25 nt between the EcoRI and HindIII sites. (2) A physical map of a piece of DNA showing recognition sites of specific restriction endonucleases separated by lengths marked in numbers of bases.

restriction site. See Restriction enzyme.

restriction. To "restrict" DNA means to cut it with a restriction enzyme. See "Restriction Enzyme".

restrictive condition. An environmental condition under which a conditional mutant shows the mutant phenotype.

restrictive temperature. A temperature at which temperature-sensitive mutants display the mutant phenotype.

retinoblastoma. A cancer of the human retina. Predisposition to retinoblastoma is inherited as an autosomal dominant. A childhood cancer of retinoblast cells caused by the inactivation of an anti-oncogene.

retrotransposon (retroposon). A class of genetic elements that includes retroviruses and transposons that have an intermediate RNA stage. A transposon that was created by reverse transcription of an RNA molecule.

retrovirus. An RNA virus that replicates by first being converted into double-stranded DNA by reverse transcriptase.

reverse genetics. The experimental procedure that begins with a cloned segment of DNA, or a protein sequence, and uses this knowledge to introduce programmed mutations (through directed mutagenesis) back into the genome in order to investigate gene and protein function.

reverse transcriptase. An enzyme which will make a DNA copy of an RNA template - a DNA-dependant RNA polymerase. RT is used to make cDNA; one begins by isolating polyadenylated mRNA, providing oligo-dT as a primer, and adding nucleotide triphosphates and RT to copy the RNA into cDNA.

reverse transcriptase. An enzyme, requiring a DNA primer, that catalyzes the synthesis of a DNA strand from an RNA template. An enzyme that can use RNA as a template to synthesize DNA.

reverse transcriptase. RNA-dependent DNA polymerase.

reversion. The production of a wild-type gene from a mutant gene. The return of a mutant to the wild-type phenotype by way of a second mutational event.

RF. See recombinant frequency.

RFLP (restriction fragment length polymorphism). (1) Genetic polymorphism as revealed by the sizes of fragments generated with a particular restriction endonuclease enzyme (such as *EcoR*I, *Pst*I, *Bgl*II). (2) Restriction fragment length polymorphism; the acronym is pronounced "riflip". Although two individuals of the same species have almost identical genomes, they will always differ at a few nucleotides. Some of these differences will produce new restriction sites (or remove them), and thus the banding pattern seen on a genomic Southern will thus be affected. For any given probe (or gene), it is often possible to test different restriction enzymes until you find one which gives a pattern difference between two individuals - a

RFLP. The less related the individuals, the more divergent their DNA sequences are and the more likely you are to find a RFLP.

RFLP mapping. A technique in which DNA restriction fragment length polymorphisms (RFLPs) are used as reference markers for mapping in relation to known genes or other RFLP loci.

rho. A protein factor required to recognize certain transcription termination signals in Escherichia coli.

rho-dependent terminator. A DNA sequence signaling the termination of transcription; termination requires the presence of the rho protein.

rho-independent terminator. A DNA sequence signaling the termination of transcription; the rho protein is not required for termination.

ribonuclease. see "RNAse".

ribonucleic acid. See RNA.

riboprobe. A strand of RNA synthesized *in-vitro* (usually radiolabeled) and used as a probe for hybridization reactions. An RNA probe can be synthesized at very high specific activity, is single stranded (and therefore will not self anneal), and can be used for very sensitive detection of DNA or RNA.

ribosomal RNA (rRNA). One of the four types of RNA that exist in a eukaryotic cell. It has a very slow mutation rate which is useful in phylogenetic analysis of kingdoms and phyla.

ribosome. (1) A cellular particle which is involved in the translation of mRNAs to make proteins. Ribosomes are a complex consisting of ribosomal RNAs (rRNA) and several proteins. (2) A complex organelle (composed of proteins plus rRNA) that catalyzes translation of messenger RNA into an amino acid sequence. Ribosomes are made up of two non-identical subunits each consisting of a different rRNA and a different set of proteins. (3) A small cytoplasmic organelle that is the site of mRNA translation, thus protein synthesis.

ribosyme. RNA molecules with enzymatic activity. Its presence in organelles from plants, yeast, viruses and eukaryotic cells revolutionized the ideas about the origin of life.

ribozyme. Catalytic or autocatalytic RNA. RNA with enzymatic activity, for instance, self-splicing RNA molecules in Tetrahymena.

RNA (ribonucleic acid). (1) A single-stranded nucleic acid similar to DNA but having ribose sugar rather than deoxyribose sugar and uracil rather than thymine as one of the pyrimidine bases. (2) A single-stranded nucleic acid that is found both in nucleus and cytoplasm. Other differences from DNA are: it contains uracil instead of thymine, it is single-stranded, and its sugar molecule is ribose. Total cellular RNA is made up of ribosomal RNA (rRNA, 80-85%), transfer RNA (tRNA, 15-20%) and messenger RNA (mRNA, 1-5%). See also small nuclear RNA and heterogeneous nuclear RNA.

RNA editing. The insertion of uridines into mRNAs after transcription is completed; controlled by guide RNA (gRNA). May also sometimes involve insertion of cytidines and possible deletions of bases.

RNA *in situ* hybridization. A technique that is used to identify the spatial pattern of expression of a particular transcript (usually an mRNA). In this technique, the probe is labeled, either radioactively or by chemically attaching a fluorochrome visualised by fluorescence or an enzyme that can convert a substrate to a visible dye. A tissue or organism is soaked in a solution of single-stranded probe under conditions that allow the probe to hybridize to complementary RNA sequences in the cells; unhybridized probe is then removed. Radioactive probe is detected by autoradiography. Fluorochrome is detected by fluorescence microscopy. Enzyme labeled probe is detected by soaking the tissue in the substrate; the dye develops in sites where the transcript of interest was expressed.

RNA phage. Phage whose genetic material is RNA. They are the simplest known phages.

RNA polymerase. (1) An enzyme that catalyzes the synthesis of an RNA strand from a DNA template. The enzyme that polymerizes RNA by using DNA as a template. It can also act as a primase initiating DNA replication. (Also known as transcriptase or RNA transcriptase). (2) An enzyme that transcribes an RNA molecule from the template strand of a DNA molecule. It adds to the 3' end of the growing RNA molecule one nucleotide at a time using ribonucleotide triphosphates (rNTPs) as substrates (this reaction releases pyrophosphates). RNA polymerase I is dedicated to the synthesis of only one type of RNA molecule (pre-rRNA). RNA polymerase II is required for general transcription reactions. RNA polymerase III produces small RNAs such as tRNAs and 5S rRNA.

RNA replicase. A polymerase enzyme that catalyzes the self-replication of single-stranded RNA.

RNase protection assay. This is a sensitive method to determine (1) the amount of a specific mRNA present in a complex mixture of mRNA and/or (2) the sizes of exons which comprise the mRNA of interest. A radioactive DNA or RNA probe (in excess) is allowed to hybridize with a sample of mRNA (for example, total mRNA isolated from tissue), after which the mixture is digested with single-strand specific nuclease. Only the probe which is hybridized to the specific mRNA will escape the nuclease treatment, and can be detected on a gel. The amount of radioactivity which was protected from nuclease is proportional to the amount of mRNA to which it hybridized. If the probe included both intron and exons, only the exons will be protected from nuclease and their sizes can be ascertained on the gel.

RNase. Ribonuclease; an enzyme which degrades RNA. It is ubiquitous in living organisms and is exceptionally stable. The prevention of RNase activity is the primary problem in handling RNA.

Robertsonian fusion. Fusion of two acrocentric chromosomes at the centromere.

robertsonian translocation. See centric fusion.

rolling-circle replication. A model of DNA replication that accounts for a circular DNA molecule producing linear daughter double helices.

rotifer. Microscopic aquatic organisms (pseudocoelomates). Their peculiar features are ancient asexuality, parthenogenetic reproduction and dormancy. Despite being diploid, they don't have homologous chromosomes.

rRNA (ribosomal RNA). (1) A class of RNA molecules, coded in the nucleolar organizer, that have an integral (but poorly understood) role in ribosome structure and function. RNA components of the subunits of the ribosomes. (2) "Ribosomal RNA"; any of several RNAs which become part of the ribosome, and thus are involved in translating mRNA and synthesizing proteins. They are the most abundant RNA in the cell (on a mass basis).

RT-PCR. See 'Polymerase Chain Reaction'.

run-off. See Nuclear run-on.

run-on. See Nuclear run-on.

S

S (Svedberg unit). A unit of sedimentation velocity, commonly used to describe molecular units of various sizes (because sedimentation velocity is related to size).

S1 end mapping. A technique to determine where the end of an RNA transcript lies with respect to its template DNA (the gene). Can't be described in a short paragraph. See "RNAse Protection assay" for a closely related technique.

S1 nuclease. An enzyme which digests only single-stranded nucleic acids.

S-9 mix. A liver-derived cell extract used in the Ames test to activate or inactivate chemicals in a test for mutagens and potential carcinogens.

saccharomycetes cerevisiae. Unicellular Ascomycete yeast known as the baker's or brewer's yeast. Widely used as a simple eukaryotic model, particularly in recombinant DNA and cell cycle studies as well as in mating type and heterokaryon compatibility studies. It has most advantages of a prokaryotic system but is a true eukaryote. It is considered as the E.coli of the eukaryotes. S.cerevisiae can reproduce both asexually and sexually, and can be cultured in either the haploid or the diploid state. One major advantage of yeast is the ease with which specific gene disruptions, gene replacements, and gene retrievals can be accomplished.

Its complete genome was sequenced in 1997 and contains 12,057,500 bp, 6,000 genes in 16 chromosomes. It is used in the creation of YACs.

SAGE (serial analysis of gene expression). A high-throughput method that uses 10-14 bp-long tags from each cDNA expressed in a cell. The concentration of each tag sequence is proportional to the level of its mRNA in the original sample. This method is used to explore gene regulation in cell populations.

sampling distribution. The distribution of frequencies with which various possible events could occur or a probability distribution defined by a particular mathematical expression .

SAR. Scaffold attachment region - the position along eukaryotic DNA where it is anchored to the central scaffold of the chromosome.

sarcoma. Tumour of mesodermal origin (e.g., muscle, bone, cartilage).

satellite chromosome. Chromosome that seems to be an addition to the normal genome.

satellite DNA. DNA that forms a separate band in a bouyant density gradient because of its different nucleotide composition (A:T rich DNAs are less dense than G:C rich DNAs). Highly repetitive eukaryotic DNA primarily located around centromeres. Satellite DNA usually has a different buoyant density than the rest of the cell's DNA.

satellite. A terminal section of a chromosome, separated from the main body of the chromosome by a narrow constriction.

saturation mutagenesis. Induction and recovery of large numbers of mutations in one area of a genome, or in one biological function, in the hope of identifying all the genes in that area, or affecting that function.

scaffold. The eukaryotic chromosome structure remaining when DNA and histones have been removed; made from nonhistone proteins. The central framework of a chromosome to which the DNA solenoid is attached as loops; composed largely of topoisomerase.

scanning hypothesis. Proposed mechanism by which the eukaryotic ribosome recognizes the initiation region of an mRNA after binding the 5' capped end of it. The ribosome scans the mRNA for the initiation codon.

schizophyllum commune. A fungus species of Basidiomycetes group. It has thousands of mating types in a multiallelic locus, a pheromone receptor system and a pheromone system. Because of this, it has the maximal outbreeding rates (98.8%) in nature.

scientific method. A procedure used by scientists to test hypotheses by making predictions about the outcome of an experiment before the experiment is performed. The results provide support or refutation of the hypothesis.

screening technique. A technique to determine the genotype or phenotype of an organism.

screening. To screen a library (see "Library") is to select and isolate individual clones out of the mixture of clones. For example, if you needed a cDNA clone of the pituitary glycoprotein hormone alpha subunit, you would need to make (or buy) a pituitary cDNA library, then screen that library in order to detect and isolate those few bacteria carrying alpha subunit cDNA.

There are two methods of screening which are particularly worth describing: screening by hybridization, and screening by antibody.

Screening by hybridization involves spreading the mixture of bacteria out on a dozen or so agar plates to grow several ten thousand isolated colonies.

Membranes are laid onto each plate, and some of the bacteria from each colony stick, producing replicas of each colony in their original growth position). The membranes are lifted and the adherent bacteria are lysed, then hybridized to a radioactive piece of alpha DNA (the source of which is a story in itself - see "Probe"). When X-ray film is laid on the filter, only colonies carrying alpha sequences will "light up". Their position on the membranes show where they grew on the original plates, so you now can go back to the original plate (where the remnants of the colonies are still alive), pick the colony off the plate and grow it up. You now have an unlimited source of alpha cDNA.

Screening by antibody is an option if the bacteria and plasmid are designed to express proteins from the cDNA inserts (see "Expression clones"). The principle is similar to hybridization, in that you lift replica filters from bacterial plates, but then you use the antibody (perhaps generated after olde tyme protein purification rituals) to show which colony expresses the desired protein.

sea Urchin. A small spiny marine invertebrate belonging to the phylum Echinodermata. It is a model animal for the study of fertilization and development. Because it is a spawner, its gametes can be obtained in large quantities. Simple mixing of sperms and eggs causes synchronous mitosis and cytokinesis. Furthermore, its eggs are large and clear.

second division segregation (SDS). The allele arrangement (2+2+2+2) in the spores of Ascomycetes with ordered spores that indicates a crossover between a locus and its centromere. A pattern of ascospore genotypes for a gene pair showing that the two alleles segregate into different nuclei only at the second meiotic division, as a result of a single crossover between that gene pair and its centromere; can only be detected in a ordered ascus. See First division segregation

secondary oocyte. A cell formed by meiosis I in female animals.

secondary sexual character. The sex-associated phenotype of **somatic** tissues in sexually dimorphic animals.

secondary spermatocyte. The product of the first meiotic division in male animals.

secondary structure. The alpha-helical or beta-sheet configuration of the polypeptide backbone of a protein.

second-site mutation. The second mutation of a double mutation within a gene. In many cases, the second-site mutation suppresses the first mutation, so that the double mutant has the wild-type phenotype.

sector. An area of tissue or colony whose phenotype is detectably different from the surrounding tissue or colony phenotype.

sedimentation. The sinking of a molecule under the opposing forces of gravitation and buoyancy.

seed. A fertilized and mature ovule which will develop into a new plant if sown. It contains usually one fertilized ovum (embryo) and endosperm (or perisperm) which is the nutritive cell surrounding the embryo.

segmentation. The process by which the correct number of segments are established in a developing segmented animal.

segregation distortion. Violation of Mendel's first law which results in unequal segregation of a pair of alleles.

segregation, rule of. Mendel's first principle describing how genes are passed from one generation to the next.

segregation. (1) Cytologically, the separation of homologous chromosomes into different cells at cell division. (2) Genetically, the production of two separate phenotypes, corresponding to two alleles of a gene, either in different individuals (meiotic segregation) or in different tissues (mitotic segregation). (3) The separation of members of a gene pair from each other during gamete formation.

segregational load. Genetic load caused when a population is segregating less fit homozygotes because of heterozygote advantage.

segregational petite. A petite that in a cross with wild-type produces 50% petite and 50% wild-type progeny; caused by a nuclear mutation.

selection coefficient (s). The proportional excess or deficiency of fitness of one genotype in relation to another genotype.

selection Differential (S) and Response to Selection (R). Following a change in the environment, in the parental (first) generation, the mean value for the character among those individuals that survive to reproduce differs from the mean value for the whole population by a value of (S). In the second, offspring generation, the mean value for the character differs from that in the parental population by a value of R which is smaller than S. Thus, strong selection of this kind (directional) leads to reduced variability in the population. Link to a lecture on selection.

selection differential. The difference between the mean of a population and the mean of the individuals selected to be parents of the next generation.

selection progress. The difference between the mean of a population and the mean of the offspring in the next generation born to selected parents.

selection. (1) Genetic methods for obtaining desired strains often employ selective procedures based on differential survival under particular nutritional or envirnomental conditions. See conditional mutation. (2) Genetic breeding methods start by selecting particular desirable phenotypes as parents for the next generation. (3) See also natural selection.

selection-mutation equilibrium. An equilibrium allele frequency resulting from the balance between selection against the allele and mutation recreating this allele.

selective medium. A culture medium that is enriched with a particular substance to allow the growth of particular strains of organisms.

selective neutrality. A situation in which different alleles of a certain gene confer equal fitness.

selective system. An experimental technique that enhances the recovery of specific (usually rare) genotypes.

self. The fusion of male and female gametes from the same individual. See self-fertilization.

self-assembly. The ability of certain multimeric biological structures to assemble from their component parts through random movements of the molecules and formation of weak chemical bonds between surfaces with complementary shapes.

selfed. See self-fertilization.

self-fertilization (also selfing, self-pollination). (1) The fusion of male and female gametes produced by the same (hermaphrodite or bisexual) individual. Self-fertilization allows an individual to create a local population, but it fails to provide variability within a population and limits the possibilities for adaptation to environmental change. Some plants reproduce by self-fertilization but most hermaphroditic animals rarely use self-fertilization, since many of them have adaptations encouraging cross-fertilization. (2) Fertilization in which the two gametes are from the same individual.

self-incompatibility system (SI system). The genetic complex of plants that prevents self-fertilization. See also MHC and mating types.

selfish DNA. A segment of the genome with no apparent function other than to ensure its own replication.

semiconservative replication. The mode by which DNA replicates. Each strand acts as a template for a new double helix. The established model of DNA replication in which each double-stranded molecule is composed of one parental strand and one newly polymerized strand. See template.

semisterility (half sterility). The phenotype of individuals heterozygotic for certain types of chromosome aberration; expressed as a reduced number of viable gametes and hence reduced fertility. Nonviability of a proportion of gametes or zygotes.

sense mutation. A mutation that changes a termination (stop) codon into one that codes for an amino acid. Such a mutation results in an elongated protein.

sense strand. A gene has two strands: the sense strand and the anti-sense strand. *The Sense strand is, by definition, the same 'sense' as the mRNA; that is it can be translated exactly as the mRNA sequence can.* After all, the RNA must base-pair with its template strand, so the process of transcription produces the complement of the anti-sense strand. This introduces some confusion about terminology:

Some people use the term 'coding strand' and 'non-coding strand' to refer to the sense and antisense strands, respectively. Unfortunately, many people interpret these terms in exactly the opposite way. I consider the terms 'coding strand' and 'non-coding strand' to be too ambiguous. Some people use the exact opposite definition for 'sense' and 'anti-sense' that I have given here. Be aware of the possibility of a discrepancy. Textbooks I have consulted generally agree with the nomenclature given herein, albeit some avoid defining these terms at all.

sense strand. See coding strand.

Sequence tagged site (STS). Any site in a chromosome or genome that is identified by a known unique DNA sequence. STSs can be used to form genetic maps by standard mapping procedures.

sequence. As a noun, the sequence of a DNA is a buzz word for the structure of a DNA molecule, in terms of the sequence of bases it contains. As a verb, "to sequence" is to determine the structure of a piece of DNA; i.e. the sequence of nucleotides it contains.

Ser. Serine (an amino acid).

serological typing. Identification of MHC molecules expressed on cells using either naturally occurring antibodies in multiparous women or by alloantiserum raised in animals.

sex chromosome. A chromosome whose presence or absence is correlated with the sex of the bearer; a chromosome that plays a role in sex determination. Heteromorphic (different shaped, e.g., X and Y) chromosomes whose distribution in a zygote determines the sex of the organism.

sex determination. The genetic or environmental process by which the sex of an individual is established.

sex linkage. The location of a gene on a sex chromosome.

sex linked. The inheritance pattern of loci located on the sex chromosomes (usually the X chromosome in XY species); also refers to the loci themselves.

sex reversal. A syndrome known in humans and mice in which chromosomally XX individuals develop as males. In some cases, sex reversal is now known to be due to the translocation of the testis-determining region of the Y chromosome to the tip of the X chromosome in such individuals.

sex switch. A gene, normally found on the Y chromosome in mammals, that directs the indeterminate gonads toward development as testes. See testis-determining factor.

sex. Formation of new organism containing genetic material from more than a single parent.

sex-controlled trait. Traits that appear more of time in one sex than in the other.

sex-determining region Y (SRY). The sex switch or testis-determining factor in human beings located on the Y chromosome. (Sry in mice.)

sexduction. Sexual transmission of donor Escherichia coli chromosomal genes on the fertility factor. A process whereby a bacterium gains access to and incorporates foreign DNA brought in by a modified F factor during conjugation.

sex-influenced dominance. A dominant expression that depends on the sex of the individual. For example horns in sheep are dominant in males and recessive in females.

sex-influenced trait. See sex-controlled trait.

sex-lethal. A gene in Drosophila, located on the X chromosome, that is a sex switch directing development toward femaleness when it is in the on state. It is regulated by numerator and denominator elements that act to influence the genic balance equation (X/A).

sex-limited trait. Trait expressed in only one sex. It may be controlled by sex linked or autosomal loci.

sex-ratio phenotype. A trait in Drosophila whereby females produce mostly if not only daughters.

sexual dimorphism. The existence, within a species, of differences in morphology between the sexes. Examples are greater size in males of gorilla, baboon and elephant seals.

sexual reproduction. Reproduction requiring the union of sex cells (gametes) which are themselves products of meiotic division. Each offspring has a unique genetic composition due to independent assortment of chromosomes during meiosis, recombination and union of gametes.

sexual selection. (1) Natural selection operating on factors that contribute to an organism's mating success. Described by Darwin as natural selection in relation to sex. (2) The forces determined by mate choice acting to cause one genotype to mate more frequently than another genotype.

sexual spore. See spore.

sexually antagonistic. Having opposite effects in the two sexes.

Shine-Dalgarno (S-D) sequence. An eight nucleotide consensus sequence 5' UAAGGAGG 3' found in bacterial mRNAs five to ten bases before the translation initiation codon (AUG). It is thought to be involved in initiation of translation by helping the mRNA bind to the ribosome (16S rRNA), thus it can be called the ribosomal binding site (see also Kozak sequence). In eukaryotic DNA, there is no such sequence. The 5' cap present on all eukaryotic mRNAs seems to be the first signal to start protein synthesis.

Shine-Delgarno hypothesis. A proposal that prokaryotic mRNA is aligned at the ribosome by complementarity between the mRNA upstream from the initiation codon and the 3' end of the 16S rRNA.

shotgun cloning. The practice of randomly clipping a larger DNA fragment into various smaller pieces, cloning everything, and then studying the resulting individual clones to figure out what happened. For example, if one was studying a 50 kb gene, it "may" be a bit difficult to figure out the restriction map. By randomly breaking it into smaller fragments and mapping those, a master restriction map could be deduced. See also Shotgun sequencing.

shotgun sequencing. A way of determining the sequence of a large DNA fragment which requires little brainpower but lots of late nights. The large fragment is shotgun cloned (see above), and then each of the resulting smaller clones ("subclones") is sequenced. By finding out where the subclones overlap, the sequence of the larger piece becomes apparent. Note that some of the regions will get sequenced several times just by chance.

shotgun technique. Cloning a large population of different DNA fragments, known to contain a fragment of interest, as a prelude to selecting or screening for that one particular clone containing the fragment of interest for intensive study.

shuttle vector. A vector (e.g. a plasmid) constructed in such a way that it can replicate in at least two different host species (e.g., a prokaryote and a eukaryote). A DNA recombined into such a vector can be tested or manipulated in several cell types.

sibling species. Two species evolved from a common ancestor and are genetically distinct but morphologically similar.

siblings(sibs). Brothers and sisters.

sickle-cell anemia. Potentially lethal human disease, inherited as an autosomal recessive, caused by a mutation in a gene coding for the beta subunit of the oxygen-transporting protein haemoglobin. Under conditions of low oxygen tension, the altered beta-globin molecule causes haemoglobin to aggregate forming rod-like arrays that distort the cell membrane forcing red blood cells to become sickle-shaped. The sickled red blood cells are damaged and rapidly removed from the circulation causing anaemia.

sigma factor. The protein that gives promoter-recognition specificity to the RNA polymerase core enzyme of bacteria.

signal hypothesis. The major mechanism whereby proteins that insert into or cross a membrane are synthesized by a membrane-bound ribosome. The first thirteen to thirty-six amino acids synthesized, termed a signal peptide, are recognized by a signal recognition particle that draws the ribosome to the membrane surface by interaction with a docking protein. The signal peptide may later be removed from the protein.

signal peptide. See signal hypothesis.

signal recognition particle. See signal hypothesis.

signal sequence. A stretch of hydrophobic amino acids at the amino-terminal of a protein that guides protein translocation through cellular membranes such as lysosomal membrane. It helps the protein to pass through the membrane via interaction with its receptor on the membrane and is usually cleaved off at the final destination by an endopeptidase. Sometimes used interchangeably with leader sequence.

signal sequence. The N-terminal sequence of a secreted protein, which is required for transport through the cell membrane.

signal transduction cascade. Refers to a series of sequential events, such as protein phosphorylations, consequent upon binding of ligand by a transmembrane receptor, that transfer a signal through a series of intermediate molecules until final regulatory molecules, such as transcription factors, are modified in response to the signal.

signal transduction. A complex multistep pathway from the cell membrane through the cytoplasm to the nucleus involved in cell proliferation and differentiation. The components are growth factors, growth factor receptors, membrane and cytoplasmic tyrosine kinases, GTP-binding (G) proteins, nuclear binding proteins and transcription factors.

silencer. A DNA sequence which acts in the opposite direction of an enhancer to inhibit the transcription of a gene.

silent mutation. (1) Base-pair substitution which alters a codon but does not result in altered phenotype due to the degeneracy of the genetic code (synonymous mutation). (2) Mutation in which the function of the protein product of the gene is unaltered.

SINE. (1) Short interspersed element. A type of small dispersed repetitive DNA sequence (e.g., Alu family in the human genome) found throughout a eukaryotic genome. (2) An abundant intermediate DNA sequence in mammals about 300 bp long (see also LINES).

Single Nucleotide Polymorphism (SNP). A single nucleotide change in the DNA code. It is the most common type of stable genetic variation and usually bi-allelic. SNPs may be silent -no change in phenotype- (sSNP), may cause a change in phenotype (cSNP) or may be in a regulatory region (rSNP) with potential to change phenotype. Anonymous SNPs are the most common ones. These are in non-coding regions and used as genetic markers. On average, each 1 kb of human genome contains 2-10 SNPs, i.e., one in every 100-500 nucleotides is polymorphic most frequently a C to T substitution.

single-strand binding protein. Protein that binds to single-stranded DNA usually near the replication fork to stabilize the single strands.

sister chromatid exchange. An exchange (crossing-over) of genetic material between the two (identical) chromatids of a chromosome in mitosis (mitotic recombination). Normally, genetic recombination takes place between homologous chromosomes in meiosis I. Sister chromatid exchange may be a sign of chromosomal instability but has no genetic consequences as long as the exchange is the result of an equal crossover.

sister chromatids. See chromatids.

sister-chromatid exchange (SCE). An event similar to crossing over that can occur between sister chromatids at mitosis or at meiosis.

site-specific recombination. A crossover event, such as the integration of phage lambda, that requires homology of only a very short region and uses an enzyme specific for that recombination. Recombination occurring between two specific sequences that need not be homologous; mediated by a specific recombination system.

skew. A distortion of the shape of the normal distribution toward one side or the other.

slot blot. Similar to a dot blot, but the analyte is put onto the membrane using a slot-shaped template. The template produces a consistently shaped spot, thus decreasing errors and improving the accuracy of the analysis. See Dot blot.

small nuclear RNA (snRNA). Small (90 to 300 nucleotides) RNA molecules that are not directly involved in protein synthesis but may have roles in RNA processing (splicing) and the cellular architecture.

snRNA. Small nuclear RNA; forms complexes with proteins to form snRNPs; involved in RNA splicing, polyadenylation reactions, other unknown functions (probably).

snRNP. (1) "Snerps", Small Nuclear RiboNucleoProtein particles, which are complexes between small nuclear RNAs and proteins, and which are involved in RNA splicing and polyadenylation reactions. (2) Small nuclear ribonucleoprotein (RNA plus protein) particle. Component of the spliceosome, the intron-removing apparatus in eukaryotic nuclei .

sociobiology. The study of the evolution of social behavior in animals.

solenoid structure. The supercoiled arrangement of DNA in eukaryotic nuclear chromosomes produced by coiling the continuous string of nucleosomes (about 7 nucleosomes per turn).

solution hybridization. A method closely related to RNase protection (see "RNase protection assay"). Solution hybridization is designed to measure the levels of a specific mRNA species in a complex population of RNA. An excess of radioactive probe is allowed to hybridize to the RNA, then single-strand specific nuclease is used to destroy the remaining unhybridized probe and RNA. The "protected" probe is separated from the degraded fragments, and the amount of radioactivity in it is proportional to the amount of mRNA in the sample which was capable of hybridization. This can be a very sensitive detection method.

somatic cell. A cell that is not destined to become a gamete; a cell whose genes will cannot be passed on to future generations.

somatic doubling. A disruption of the mitotic process that produces a cell with twice the normal chromosome number.

somatic hypermutation. The occurrence of a high level of mutation in the variable regions of immunoglobulin genes.

somatic mutation. A mutation occurring in a somatic cell.

somatic recombination. Rearrangement of genes in cells other than germ cells which happens to generate the extreme diversity of T-cell receptors and immunoglobulins (see adaptive immunity).

somatic-cell genetics. Asexual genetics, involving study of somatic mutation, mitotic crossing-overand segregation.

somatic-mutation theory. A theory to account for the high degree of antibody variability. It suggests that mutation of a basic immunoglobulin gene accounts for all the different types of immunoglobulins produced by B lymphocytes. See germ-line theory.

somatostatin. A human growth hormone.

SOS box. The region in the promoter of various genes that is recognized by the LexA repressor. Release of repression results in the induction of the SOS response.

SOS repair. The error-prone process whereby gross structural DNA damage is circumvented by allowing replication to proceed past the damage through imprecise polymerization.

SOS response. Repair systems (recA; uvr) induced by the presence of single-stranded DNA that usually occurs from postreplicative gaps caused by various types of DNA damage. The RecA protein, stimulated by single-stranded DNA, is involved in the inactivation of the LexA repressor thereby inducing the response.

southern blot. (1) A technique for analyzing mixtures of DNA, whereby the presence and rough size of one particular fragment of DNA can be ascertained. See "Blotting". Named for its inventor, Dr E. M. Southern. (2) Transfer of electrophoretically separated fragments of DNA, after denaturation, from the gel to an absorbent sheet of material, such as nitrocellulose, to which the DNA binds. The sheet is immersed in a solution containing a labeled probe that will hybridize to fragment(s) of interest. The method was first devised by E. M. Southern to transfer DNA fragments from an agarose gel to a nitrocellulose paper for hybridization, but similar transfer methods are now also used for transfering RNA or protein to papers of a variety of types followed by hybridization (RNA) or labeled antibody treatment (protein) to identify specific molecules.

spacer DNA. Regions of non-transcribed DNA between transcribed repeated genes such as ribosomal RNA genes in eukaryotes. Its function is probably to do with ensuring the high rates of transcription associated with these genes.

specialized (restricted) transduction. The situation in which a particular phage will transduce only specific regions of the bacterial chromosome. Form of transduction based on faulty looping out by a temperate phage. Only loci neighbouring the attachment site can be transduced. See generalized transduction.

speciation. A process whereby over time one species evolves into a different species (anagenesis) or whereby one species diverges to become two or more species (cladogenesis).

Speciation. It is now almost universally agreed that the prevailing process of speciation is geographical (allopatric) speciation. (There are also parapatric and sympatric speciation concepts.) According to biological species concept, however, species are defined as aggregation of populations that are reproductively isolated from one another.

species. (1) A group of individuals which can successfully breed with each other to produce offspring which can still breed with each other. There are evolutionary, biological, and recognition species concepts. (2) A group of organisms belong to the same biological species if they are capable of interbreeding to produce fertile offspring. However the biological test of a species is not always available, and so there is also a morphological species concept based on anatomical similarities.

specific-locus test. A system for detecting recessive mutations in diploids. Normal individuals treated with mutagen are mated to testers that are homozygous for the recessive alleles at a number of specific loci; the progeny are then screened for recessive phenotypes.

sperm cell. The male gamete.

sperm competition. Competition not for access to females but for fertilization of egg. Equivalent to pollen tube competition in plants and a type of sexual selection. See Birkhead's book *Promiscuity* (2000) for a detailed study of sperm competition in nature.

spermatid. The four products of meiosis in males that develop into sperm.

spermatocyte. (1) (Zoological) A cell which undergoes meiosis and thereby produces spermatids. See primary spermatocyte and secondary spermatocyte. (2) (Botanical) A cell that becomes converted to a spermatozoid (a small motile male gamete with flagella) without intervention of cell division.

spermatocyte. A male germ cell that undergoes meiosis and produces a haploid spermatid and subsequently a sperm.

spermatogenesis. The process of sperm production.

spermatogonium. A cell type in the testes of male vertebrates that gives rise to primary spermatocytes by mitosis.

spermatophyta (sperm plants; Gr. sperm=seed; phyton=plant). A major subdivision within the vascular plants division of the plant Kingdom characterized by reproduction by seeds. Sperm plants include angiosperms and gymnosperms.

spermiogenesis. The process by which spermatids mature into sperm cells.

spindle. The set of tubulin fibers that appear to move eukaryotic chromosomes during division. The microtubule apparatus that controls chromosome movement during mitosis and meiosis.

spiral cleavage. The cleavage process in molluscs and some invertebrates whereby orientation of the spindle at mitosis is at an angle to the original egg axis.

spirillum. A spiral bacterium.

spliceosome. Protein-RNA complex that removes introns from eukaryotic nuclear RNAs.

splicing. An event which takes place within the nucleus whereby introns are removed from the precursor mRNA and the exons are joined together as a post-transcriptional modification.

splicing. The reaction that removes introns and joins together exons in eukaryotic nuclear primary RNA transcripts.

spontaneous mutation. A mutation occurring in the absence of mutagens, usually due to errors in the normal functioning of cellular enzymes.

spore. (1) In plants and fungi, sexual spores are the haploid cells produced by meiosis. (2) In fungi, asexual spores are somatic cells that are cast off to act either as gametes or as the initial cells for new haploid individuals. (3) In plants with alternation of generations, small reproductive bodies capable of giving rise to a new offspring either immediately or after a period of dormancy. It can be produced asexually or sexually. It usually germinates without fusing with another cell. Sexual spores of plants are haploid cells produced by meiosis.

sporophyte. (1) The diploid (asexual) spore-producing generation in plants with alternation of generations. A sporophyta is typically formed by the union of sexual cells produced by the gametophyta. In higher plants, the sporophyte is the conspicuous plant. In lower plants (such as mosses), the gametophyte is the dominant generation. (2) The diploid sexual-spore-producing generation in the life cycle of plants. The stage in which meiosis occurs. The stage of a plant life cycle that produces spores by meiosis and alternates with the gametophyte stage.

SSOP. Sequence-specific oligonucleotide probe. Together with PCR-SSP, commonly used to type classical MHC genes following amplification by PCR reaction (see also serological typing).

SSP. Sequence-specific primer. PCR-SSP is a common method to type classical MHC genes.

stabilizing selection. (1) A type of selection that removes individuals from both ends of a phenotypic distribution thus maintaining the same distribution mean. (2) Natural selection against extreme deviations from the average (like low and high birth weight).

stable transfection. A form of transfection experiment designed to produce permanent lines of cultured cells with a new gene inserted into their genome. Usually this is done by linking the desired gene with a "selectable" gene, i.e. a gene which confers resistance to a toxin (like G418, aka Geneticin). Upon putting the toxin into the culture medium, only those cells which incorporate the resistance gene will survive, and essentially all of those will also have incorporated the experimenter's gene.

stacking. The packing of the flattish nitrogen base-pairs at the centre of the DNA double helix.

staggered cuts. The cleavage of two opposite strands of duplex DNA at points near one another.

standard deviation. The square root of the variance.

standard error of the mean. The standard deviation divided by the square root of the sample size. It is the standard deviation of a sample of means.

stasipatric speciation. Instantaneous speciation caused by polyploidy.

statistic. A computed quantity characteristic of a population, such as the mean.

statistical distribution. The array of frequencies of different quantitative or qualitative classes in a population.

superantigen. An antigen of a virus or bacteria which bind to non-polymorphic parts (outside of the antigen-binding cleft) of MHC class II molecules and interact with the V? domain of the T-cell receptor. This way, they activate an entire subgroup of T cells (rather than a specific clone) expressing the appropriate V? and this is followed by deletion of the activated T cells upon exposure to the superantigen. One superantigen can activate up to 20% of the helper T cell repertoire. The prototype bacterial superantigen is the staphylococcal enterotoxins.

supercoil. A closed double-stranded DNA molecule that is twisted on itself.

supercoiled/supertwisted DNA. A closed circular DNA molecule in which the DNA molecule is further twisted on itself to form a more compact molecule. Left-handed (negative) supercoiling leads to a loosening of the strands of the double helix (underwinding). Positive supercoiling is not seen in vivo.

supercoiling. Negative (tending to unwind the helix) or positive (tending to wind the helix) coiling of double-stranded DNA that differs from the relaxed state.

supergene. Several loci, which usually control related aspects of the phenotype, in close physical association.

superinfection. Phage infection of a cell that already harbours a prophage.

supersuppressor. A mutation that can suppress a variety of other mutations; typically a nonsense suppressor.

suppressive petite. A petite that in a cross with wild-type produces progeny of which variable non-Mendelian proportions are petite.

suppressor gene. A gene that, when mutated, apparently restores the wild-type phenotype to a mutation at another locus.

suppressor mutation. A mutation that counteracts the effects of another mutation. A suppressor maps at a different site than the mutation it counteracts, either within the same gene or at a more distant locus. Different suppressors act in different ways.

survival of the fittest. In evolutionary theory the survival of only those organisms best able (fittest) to obtain and utilize resources. This phenomenon is the cornerstone of Darwin's theory of evolution by natural selection.

Svedberg unit. A unit of sedimentation rate during centrifugation. Abbreviation is S as in 50S.

swivelase. See DNA gyrase.

symbiosis. Two organisms living together and both benefit from this. An example is the coral and algae (zooxanthellae).

sympatric speciation. (1) Speciation in which the evolution of reproductive isolating mechanisms occurs within the range and habitat of the parent species. This form of speciation may be common in parasites. (2) Speciation that occurs as a result of divergence of two populations occupying the same geographical area. It is due to reproductive isolation of subpopulations of a species without physical isolation. One well-established mechanism is polyploidy.

synapsis. The point-by-point pairing of homologous chromosomes during zygotene or in certain Dipteran tissues (e.g., Drosophila salivary glands) that undergo endomitosis. Close pairing of homologs at meiosis.

synaptonemal complex. A proteinaceous complex that apparently mediates synapsis during zygotene stage and then disintegrates. A complex structure that unites homologous chromosomes during the prophase of meiosis.

syncytial blastoderm. In insects, the syncytial stage of blastoderm preceding the formation of cell membranes around the individual nuclei of the early embryo.

syncytium. A single cell with many nuclei.

syngamy. The union of the (haploid) nuclei of two gametes following fertilization to form a single (diploid) nucleus for the zygote.

syngeneic. Genetically identical (isogeneic) members of the same species like monozygotic twins.

synonymous (silent) base change. A change in the nucleotide sequence that does not cause an amino acid change. Non-synonymous changes replace the amino acid and are called replacement change.

synteny test. A test that determines whether two loci belong to the same linkage group (i.e. are syntenic) by observing concordance (occurence of markers together) in hybrid cell lines.

synteny. (1) All loci on one chromosome are said to be syntenic (literally on the same ribbon). Loci may appear to be unlinked by conventional genetic tests for linkage but still be syntenic. (2) Refers to two genomes in which certain groups of genes are conserved in similar regional maps. Parts of mouse chromosome 17 and human chromosome 6 are syntenic.

synthetic medium A chemically defined substrate upon which organisms are grown.. (1) Thymine or thymidine. (2) See tetratype.

synthetic theory of evolution. Proposed to explain the transformation of a species by natural selection and for the splitting of a species into reproductively isolated subgroups.

systematics. Classification of living things with regard to their evolutionary relationship.

T

T cells. A subgroup of T lymphocytes characterized by having T-cell receptor (TCR) complex and CD3 surface marker. T cells are roughly subdivided into CD4+ helper t cells and CD8+ cytotoxic and suppressor T cells.

TACTAAC box. A consensus sequence surrounding the lariat branch point of eukaryotic pre-mRNA introns and used during splicing.

tandem duplication. adjacent identical chromosome segments.

taq polymerase. A DNA polymerase isolated from the bacterium *Thermophilis aquaticus* and which is very stable to high temperatures. It is used in PCR procedures and high temperature sequencing.

target theory. A theory that predicts response curves based on the number of events required to cause the phenomenon. Used to determine whether point mutations are single events.

targeted gene knockout. The introduction of a null mutation in a gene by a designed alteration in a cloned DNA sequence that is then introduced into the genome by homologous recombination and replacement of the normal allele.

TATA box (Goldberg-Hogness box). (1) A short nucleotide sequence in the promoter 25 to 35 bp upstream to the transcription initiation (cap) site of eukaryotic genes to which RNA polymerase II binds. The consensus sequence is TATAA/TA. The TATA box binds the general transcription factor TFIID (see also CAAT box). (2) A DNA sequence (concensus TATAAA) at about minus 25 (ie. 25 nucleotides upstream of the transcription start site) in the promoter region of many eukaryotic genes involved in binding RNA polymerase via a TATA binding protein. Analogous to the Pribnow box in prokaryotes. (3) A sequence found in the promoter (part of the 5' flanking region) of many genes. Deletion of this site (the binding site of transcription factor TFIID) causes a marked reduction in transcription, and gives rise to heterogeneous transcription initiation sites.

tautomeric shift. The spontaneous isomerization of a nitrogen base to an alternative hydrogen-bonding form, possibly resulting in a mutation. Reversible shifts of proton position in a molecule. Bases in nucleic acids shift between keto and enol forms or between amino and imino forms.

taxon. Any group of organisms to which any rank of taxonomic name (classification) is applied. Plural: taxa.

taxonomic hierarchy. All taxa are classified within the following groups (starting from the most inclusive): kingdom, ('division' in plants), phylum, class, order, family, genus, species, subspecies (race).

TCA cycle. The tricarboxylic acid cycle (also known as the Krebs cycle). A major metabolic pathway involved in aerobic cellular respiration (energy production) which takes place in the mitochondria of animal and plant cells. Pyruvic acid produced during glycolysis is converted to acetyl CoA which is then oxidized to CO_2. The reducing power of the end products NADH and

$FADH_2$ are used in the synthesis of ATP by oxidative phosphorylation.

T-cell receptor. Surface protein of T lymphocyte cells that allow the T cells to recognize infected host cells.

t-dna. A portion of the Ti plasmid that is inserted into the genome of the host plant cell.

teleology. Teleology, from the Greek word telos (purpose), asserts that there is an element of purpose or design behind the workings of nature. Attributing any purposeful direction to evolutionary change would be called teleological.

teleost. Bony fishes with well-developed bone structure.

telocentric chromosome. A chromosome whose centromere lies at one end.

telocentric. A chromosome with a terminal centromere (like chromosome 21 in humans).

telomerase. A **reverse transcriptase** (h*TERT*) containing an RNA molecule (h*TR*) that functions as the template for the tandem repeat at telomere. It synthesizes telomere to maintain its length after each cell division. It is active in embryonic cells and gametes, inactive in differentiated somatic cells, and reactivated in malignant cells. Telomerase can add one base at a time to the telomeric end of a chromosome. This maintenance work is required for cells to escape from replicative senescence. Telomerase activity is the most general molecular marker for identification of human cancer.

telomerase. An enzyme that adds telomeric sequences to the ends of eukaryotic chromosomes. No template is necessary.

telomere. (1) The end of eukaryotic chromosome consisting of tandemly repeated sequences. Chromosomes lose about 100 bp from telomere every time the cell divides. The enzyme telomerase can add the lost bases. (2) The ends of linear chromosomes that are required for replication and stability. The tip (or end) of a chromosome.

telophase. The terminal stage of mitosis or meiosis in which chromosomes uncoil; the spindle breaks down; and cytokinesis usually occurs. The late stage of nuclear division when daughter nuclei re-form.

temperate phage. A phage that can enter into lysogeny with its host. A phage that can become a prophage.

temperature-sensitive mutant. An organism that has a wild-type phenotype at a permissive temperature but a mutant phenotype at a restrictive (non-permissive) temperature.

temperature-sensitive mutation. A conditional mutation that produces the mutant phenotype in one (restrictive or non-permissive) temperature range and the wild-type phenotype in another (permissive) temperature range.

template strand. See anticoding strand.

template. (1) A pattern serving as a mechanical guide. (2) In DNA replication each strand of the duplex acts as a template for the synthesis of a new double helix. (3) A molecular mold that shapes the structure or sequence of another molecule. For example, the nucleotide sequence of DNA acts as a template to control the nucleotide sequence of RNA during transcription.

teratogen. An agent that interferes with normal development.

terminal redundancy. A linear DNA molecule with the same sequence (genetic information) at each end. If genetic information is represented by

ABCDEFGH

then a terminally redundant sequence can be for instance

ABCDEFGHAB.

Terminal redundancy is seen in some phages (e.g., T2) and is generated because a phage head is capable of containing a DNA molecule larger than the complete genome and packaging of DNA into phage heads is determined by the headful. These phages also show circular permutation.

terminator sequence. A sequence in DNA that signals termination of transcription to RNA polymerase.

tertiary structure. The further folding of a protein bringing alpha-helices and beta-sheets into three-dimensional arrangements. The folding or coiling of the secondary structure to form a three-dimensional molecule.

test-cross. The crossing of an organism, with an unkown genotype, to a homozygous recessive organism (tester). A cross between an individual of unknown genotype or a heterozygote (or a multiple heterozygote) to a homozygous recessive individual.

tester. An individual homozygous for one or more recessive alleles and used in a test-cross.

testicular feminization syndrome. A human condition, inherited as an X-linked recessive, caused by a mutation in a gene coding for the androgen (testosterone) receptor, in which genetical **(XY)** males develop female secondary sexual characters.

testicular feminization. An X-linked trait that causes XY individuals to develop into phenotypic females. A mutation causes loss of sensitivity to testosterone.

testing of hypotheses. The determination of whether to accept or reject a proposed hypothesis based on the likelihood of the experimental results.

testis-determining factor (TDF). General term for the gene determining maleness in human beings (Tdf in mice).

tet resistance. See “Antibiotic resistance”.

tetrad analysis. The use of tetrads (definition 2) to study the segregation of chromosomes and genes during meiosis.

tetrad. (1) Four homologous chromatids in a bundle in the first meiotic prophase and metaphase. The meiotic configuration of four chromatids first seen in pachytene. There is one tetrad per pair (bivalent) of homologous chromosomes. (2) The four haploid product cells from a single meiosis.

tetraparental mouse. A mouse that develops from an embryo created by the experimental fusion of two separate embryos (blastulas) from matings of different parents.

tetraploid. (1) A cell having four chromosome sets. (2) an organism composed of such cells.

tetrapods. Vertebrate animals other than fishes (amphibians, reptiles, birds, and mammals).

tetrapolar. Used for the mating types of Basidiomycete to describe four distinct ways of interactions between haploid mycelia. These fungi have two mating type loci and there are four degrees of matching: fully compatible at both loci, fully incompatible at both loci, semicompatible (compatible only at locus 1), and semicompatible (compatible only at locus 2). In Ascomycete, the mating type locus is biallelic and mating types are bipolar.

tetratype (T). A tetrad containing four different genotypes, two parental and two recombinant. A spore arrangement in **Ascomycete**s that consists of two parental and two recombinant spores indicating a single crossover between two linked loci. See parental ditype (PD) and non-parental ditype (NPD).

theta structure. An intermediate structure formed during the replication of a circular DNA molecule.

Thr. Threonine (an amino acid).

three-point cross. A cross involving three loci.

three-point testcross. A testcross involving one parent with three heterozygous gene pairs and another (tester) with three homozygous recessive gene pairs.

three-prime (3') end. The end of a DNA or RNA strand with a free 3' hydroxyl group corresponding to the end of transcription (see also five-prime end).

thymidine. The nucleoside having thymine as its base.

thymine dimer. A pair of abnormally chemically bonded adjacent thymine bases in DNA, resulting from damage by ultra-violet irradiation. The cellular processes that repair this lesion often make errors that create mutations.

thymine. A pyrimidine base that pairs with adenine.

Ti (tumor-inducing) plasmid. A plasmid of Agrobacterium tumefaciens that is often used as a vector in genetic engineering of plants. During tumor induction, a transposon of the plasmid (T-DNA) integrates into the host chromosome.

Ti plasmid. A circular plasmid of Agrobacterium tumefaciens that enables the bacterium to infect plant cells and produce a tumour (crown gall tumour).

tissue-specific expression. Gene function which is restricted to a particular tissue or cell type. For example, the glycoprotein hormone alpha subunit is produced only in certain cell types of the anterior pituitary and placenta, not in lungs or skin; thus expression of the glycoprotein hormone alpha-chain gene is said to be tissue-specific. Tissue specific expression is usually the result of an enhancer which is activated only in the proper cell type.

topoisomer. Topological form of DNA with the same sequence as another but differing in linking number (supercoiling).

topoisomerase. (1) A class of enzymes that convert DNA from one topological form to another. During DNA replication, they facilitate the untwisting of supertwisted DNA (see also gyrase and helicase). (2) Enzymes of two types that can remove (or create) supercoiling in duplex DNA by creating transitory breaks in one (type I) or both (type II) strands of the sugar-phosphate backbone. See topoisomer

totipotency. The ability of a cell to proceed through all the stages of development and thus produce a nórmal adult.

totipotent. The state of a cell that can give rise to any and all adult cell types as compared with a differentiated cell whose fate is determined.

trans conformation. In a heterozygote involving two mutant sites (ab) within a gene or gene cluster, the arrangement, Ab/aB. See cis conformation

trans. Meaning across and referring usually to the geometric configuration of two mutant alleles (ab) across from each other on a pair of homologous chromosomes (Ab/aB). See cis.

trans-acting gene. A gene acting on or co-operating with another gene on a different chromosome (see also cis-acting gene).

introduced into cells. The introduction of foreign DNA into eukaryotic or prokaryotic cells.

transfer operon (tra). Operon containing genes that produce the male (F-pili producing) bacterial phenotype. The F+(plus) cell (male cell) can transfer the fertility factor (F plasmid) to an F-(minus) (female) cell.

transfer RNA. See tRNA.

transform. (1) In microbiology and genetics the modification of one genotype by the external application of DNA from a cell of another genotype. (2)In oncology (the study of cancer formation) the conversion of a normal cell into a cancerous one by mutation or transfection.

transformation (with respect to bacteria). The process by which a bacteria acquires a plasmid and becomes antibiotic resistant. This term most commonly refers to a bench procedure performed by the investigator which introduces experimental plasmids into bacteria.

transformation (with respect to cultured cells). A change in cell morphology and behavior which is generally related to carcinogenesis. Transformed cells tend to exhibit characteristics known collectively as the "transformed phenotype" (rounded cell bodies, reduced attachment dependence, increased growth rate, loss of contact inhibition, etc). There are different "degrees" of transformation, and cells may exhibit only a subset of these characteristics. Not well understood, the process of transformation is the subject of intense research.

transformation. (1) The modification of a genome by the external application of DNA from a cell of different genotype. (2) Conversion of normal eukaryotic cells to a cancer-like state of uncontrolled division. (3) In bacterial transformation, it means the transfer of genes from one bacterium to another in the forms of soluble fragments of DNA; in malignant transformation, it means conversion of normal animal cell state to unregulated growth.

transformer. An allele in fruit flies (Drosophila) that converts chromosomal females into sterile males.

transgenic mouse. A mouse which carries experimentally introduced DNA. The procedure by which one makes a transgenic mouse involves the injection of DNA into a fertilized embryo at the pro-nuclear stage. The DNA is generally cloned, and may be experimentally altered. It will become incorporated into the genome of the embryo. That embryo is implanted into a foster mother, who gives birth to an animal carrying the new gene. Various experiments are then carried out to test the functionality of the inserted DNA.

transgenic organism. An organism whose genome has been modified by introduction of novel DNA.

transgenic. Organisms that have had foreign DNA stably integrated into their genome.

transient diploid. The stage of the life cycle of predominantly haploid fungi (and algae) during which meiosis occurs .

transient transfection. When DNA is transfected into cultured cells, it is able to stay in those cells for about 2-3 days, but then will be lost (unless steps are taken to ensure that it is retained - see Stable transfection). During those 2-3 days, the DNA is functional, and any functional genes it contains will be expressed. Investigators take advantage of this transient expression period to test gene function.

transition mutation. A mutation in which a purine/pyrimidine base pair is replaced with a base pair in the same purine/pyrimidine relationship e.g. GC with AT.

transition. (1) A nucleotide substitution between two purine nucleotides (A and G) or between pyrimidines (C and T/U) in DNA or RNA. Transition-type substitution is more common than transversion. (2) A type of nucleotide-pair substitution involving the replacement of a purine with another purine, or of a pyrimidine with another pyrimidine for example GC with AT. See transversion.

translation. (1) The process of converting the RNA sequence into the linear sequence of amino acids in a protein product. Translation start site contains a codon (AUG) for methionine but not all proteins start with a methionine as most of the time it is cleaved off post-translationally. (2) The process of decoding a strand of mRNA, thereby producing a protein based on the code. This process requires ribosomes (which are composed of rRNA along with various proteins) to perform the synthesis, and tRNA to bring in the amino acids. Sometimes, however, people speak of "translating" the DNA or RNA when they are merely reading the nucleotide sequence and predicting from it the sequence of the encoded protein. This might be more accurately termed "conceptual translation". (3) The process of protein synthesis whereby the primary structure of the protein is determined by the nucleotide sequence in mRNA. The ribosome-mediated production of a polypeptide whose amino acid sequence is derived from the codon sequence of an mRNA molecule.

translocase (EF-G). Elongation factor in prokaryotes necessary for proper translocation at the ribosome during the translation process. Replaced by eEF2 in eukaryotes.

translocation. 1. The relocation of a chromosomal segment in a different position in the genome. A chromosomal configuration in which part of a chromosome becomes attached to a different chromosome. 2. Also a part of the translation process in which the mRNA is shifted one codon in relation to the ribosome. 3. Transfer of chromosomal material between chromosomes (usually reciprocal).

transmembrane receptor. A protein that spans the plasma membrane of a cell, with the extracellular domain of the protein having the ability to bind to a ligand and the intracellular domain having an activity (such as a protein kinase) that can be altered (either increased or decreased) upon ligand binding.

transmission genetics. The study of the mechanisms involved in the passage of genes from one generation to the next.

transposable genetic element. A general term for any genetic unit that can insert into a chromosome, exit, and relocate; includes insertion sequences, transposons, some bacteriophages, and controlling element. A region of the genome, flanked by inverted repeats, a copy of which can be inserted at another place; also called a transposon or a jumping gene.

transposase. An enzyme that catalyzes the insertion of a **transposon**.

transposition. See transposon.

transposon. (1) A long mobile DNA element that moves in the genome by a mechanism involving DNA synthesis and transposition. (2) A mobile piece of DNA that is flanked by terminal repeat sequences and typically bears genes coding for transposition functions. See transposable genetic element.

transspecies evolution. The favored type of evolution of the MHC allelic diversity. The age of an allele or an allelic lineage is greater than the species. Therefore, common allelic lineages have been inherited from a common ancestor and species-specific mutational diversification occurred within these lineages. For example, no single MHC class I allele is shared between humans and chimpanzees, but numerous similarities in lineage, polymorphic motifs and individual substitutions can be observed. One consequence is that certain alleles will be more similar to their correspondent alleles in another species than to the other alleles of the same locus in the same species. The long-term persistence of families of MHC alleles whose origins predate speciation events is called 'trans-species evolution'.

transversion mutation. A mutation in which a purine/pyrimidine replaces a pyrimidine/purine base pair or vice versa e.g. GC with TA.

transversion. (1) A mutation caused by the substitution of a purine (A and G) for a pyrimidine (C and T/U) or vice versa in DNA or RNA (see also transition). (2) A type of nucleotide-pair substitution involving the replacement of a purine with a pyrimidine, or vice versa for example GC with TA. See transition

trihybrid. An organism heterozygous at three loci.

triplet repeat. In this situation, a triplet of nucleotides increases in number within a gene. A mutation especially occurring in central nervous system disorders is the increased number of triplets repeats. Examples include myotonic dystrophy, Huntington's disease, Friedreich's ataxia and fragile X syndrome. Also in polycystic ovary syndrome, androgen receptor gene has increased number of CAG repeats (Hickey et al, 2002). Expansion may be greater depending on the transmitting parent (e.g., the mother in myotonic dystrophy, the father in Huntington's disease); thus, a parent-of-origin effect and genetic anticipation can be observed. Increased number of repeats of a triplet may trigger methylation of the gene that causes the disease.

triplet. The three nucleotide pairs that compose a codon.

triploid. A cell having three chromosome sets, or an organism composed of such cells.

trisomic. A diploid cell with an extra chromosome. Basically a diploid with an extra chromosome of one type, producing a chromosome number of the form 2n + 1.

tritium. A radioactive isotope of hydrogen.

trna (transfer RNA). Transfer RNA. Small RNA molecules that carry amino acids to the ribosome for polymerization into a polypeptide. During translation the amino acid is inserted into the growing polypeptide chain when the anticodon of the tRNA pairs with a codon on the mRNA being translated.

tRNA. "transfer RNA"; one of a class of rather small RNAs used by the cell to carry amino acids to the enzyme complex (the ribosome) which builds proteins, using an mRNA as a guide. Fairly abundant.

Trp. Tryptophan (an amino acid).

true heritability. See heritability.

true speciation. See cladogenesis.

true-breeding line or strain. See pure-breeding line or strain.

truncation selection. A breeding technique in which individuals in whom quantitative expression of a phenotype is above or below a certain value (the truncation point) are selected as parents for the next generation.

tumor suppressor. A gene that inhibits progression towards neoplastic transformation. The best-known examples of tumor suppressors are the proteins p53 and Rb.

tumour suppressor gene. A gene encoding a protein that suppresses tumour formation. A gene that normally prevents unlimited cell division. When both copies of the gene are lost or mutated the cell is transformed to a cancer phenotype. Examples are the P53, retinoblastoma and Wilm's tumour.

tumour virus. A virus that is capable of inducing a cancer.

tumour. Abnormal growth of tissue, literally means a swelling.

Turner syndrome. An abnormal human female phenotype produced by the presence of only one X chromosome (**XO**).

twin spot. A pair of mutant sectors within wild-type tissue, produced by a mitotic crossover in an individual of appropriate heterozygous genotype.

two-point cross. A cross involving two loci.

two-strand double crossover. A double crossover that involves only two of the four chromatids of a tetrad.

type I error. In statistics the rejection of a true hypothesis.

type II error. In statistics the accepting of a false hypothesis.

typological thinking. The concept that organisms of a species conform to a specific norm. In this view variation is considered abnormal.

Tyr. Tyrosine (an amino acid).

U

U. Uracil or uridine.

ubiquitin. A small protein that becomes covalently linked to a protein targeted for degradation.

underdominance. (1) A condition in which the phenotypic expression of the heterozygote is less than that of either homozygote. (2) Also called heterozygous disadvantage. This unusual selection process occurs when heterozygotes are less fit than either homozygote. This situation is likely to arise when two adjacent populations are isolated and become homozygous for different alleles, and then come into secondary contact at the borders of their ranges. This is the opposite of overdominance.

unequal crossing over. Non-reciprocal crossing over caused by mismatching of homologous chromosomes. Usually occurs in regions of repetitive DNA.

unequal crossover. A crossover between homologs that are not perfectly aligned.

uninemic chromosome. A chromosome consisting of one double helix of DNA.

uniparental disomy. Inheritance of both homologues of a chromosome from one parent, with loss of the corresponding homologue from the other parent.

uniparental inheritance. The transmission of certain phenotypes from one parental type to all the progeny; such inheritance is generally produced by organelle (mitochondria, chloroplast) genes.

unique DNA. A length of DNA with no repetitive DNA sequences.

unmixed codon family. Group of four codons sharing their first two bases and coding for the same amino acid.

unstable mutation. A mutation that has a high frequency of reversion; a mutation caused by the insertion of a controlling element, whose subsequent exit produces a reversion.

unusual bases. Other bases in addition to the normal adenine; cytosine; guanine; and uracil. Found primarily in tRNAs and produced by post transcription modification of one of the normal bases.

upstream activator sequence. A binding site for transcription factors, generally part of a promoter region. A UAS may be found upstream of the TATA sequence (if there is one), and its function is (like an enhancer) to increase transcription. Unlike an enhancer, it can not be positioned just anywhere or in any orientation.

upstream. (1) A convention on DNA related to the position and direction of transcription by RNA polymerase (5'>3'). Downstream (or 3' to) is in the direction of transcription whereas upstream (5' to) is in the direction from which the polymerase has come. (2) Sequences located to the opposite direction to transcription (which runs from 5' to 3' on the sense strand of the DNA). A nucleotide 25 bp upstream to the first transcribed nucleotide is at position -25.

upstream/Downstream. In an RNA, anything towards the 5' end of a reference point is "upstream" of that point. This orientation reflects the direction of both the synthesis of mRNA, and its translation - from the 5' end to the 3' end. In DNA, the situation is a bit more complicated. In the vicinity of a gene (or in a cDNA), the DNA has two strands, but one strand is virtually a duplicate of the RNA, so it's 5' and 3' ends determine upstream and downstream, respectively. Note that in genomic DNA, two adjacent genes may be on different strands and thus oriented in opposite directions. Upstream or downstream is only used on conjunction with a given gene.

uracil. A pyrimidine base that appears in RNA in place of thymine found in DNA. See pyrimidine.

URF. Unassigned reading frame. An open reading frame (ORF) whose function has not yet been determined.

uridine. The nucleoside having uracil as its base.

ustilago. A genus in Basidiomycetes Phylum of the Fungal Kingdom. It represents the smut fungi that are well known for their highly polymorphic mating types.

V

Val. Valine (an amino acid).

variable expression. A variation in phenotype between affected members of the same family (i.e. individuals carrying identical mutations). It occurs in many dominant conditions and may be associated with reduced penetrance [see also penetrance].

variable region. A region in an immunoglobin molecule that shows many sequence differences between antibodies of different specificities; the part of the antibody that binds to the antigen. See immunoglobulin gene.

variable. A property that may have different values in various cases.

variable-number-of-tandem-repeats (VNTR) locus. Locus that is hypervariable because of tandemly repeated DNA sequences. Presumably variability is generated by unequal crossing over or slippage during replication. A chromosomal locus at which a particular repetitive sequence is present in different numbers in different individuals of a population or in the two different chromosome homologues in one diploid individual.

variance. The average squared deviation about the mean of a set of data. A measure of the variation around the central class of a distribution; the average squared deviation of the observations from their mean value.

variant. (1) Although commonly used to mean allele, mutant or to denote any genetic polymorphism, its precise use should be reserved for genetic markers that occur in less than 1% frequency in the population (hence cannot be called polymorphism). (2) An individual organism that is recognizably different from an arbitrary standard type in that species.

variate. A specific numerical value of a variable.

variation. The differences among parents and their offspring or among individuals in a population.

variegation. The occurrence within a tissue of sectors or clones with differing phenotypes. Patchiness; a type of position effect that results when particular loci are contiguous with heterochromatin. See position-effect variegation.

vector. (1) A plasmid, phage, or cosmid into which foreign DNA may be inserted for cloning. (2) In DNA cloning, the plasmid or phage chromosome used to carry the cloned DNA segment. (3) The DNA "vehicle" used to carry experimental DNA and to clone it. The vector provides all sequences essential for replicating the test DNA. Typical vectors include plasmids, cosmids, phages and YACs.

vertebrates. A subphylum in the Phylum **Chordata** of the Kingdom Animalia. All members have a notochord and a cranium (skull). Includes the Classes: Fishes, Amphibians, Reptiles, Birds, and Mammals (monotremes, marsupials, placentals).

viability. The probability that a fertilized egg will survive and develop into an adult organism.

virion. A virus particle.

viroid. (1) A disease-causing agent consisting of only a single-stranded, short (270 to 380 nucleotides) RNA molecule. (2) RNA molecule that is a plant pathogen.

virulent phage. A phage that cannot become a prophage; infection by such a phage always leads to lysis of the host cell.

virus. (1) A particle consisting of a nucleic acid (RNA or DNA) genome surrounded by a protein coat (capsid) and sometimes also a membrane, which can replicate only after infecting a host cell. A virus particle may exist free of its host cell but is incapable of replicating on its own. (2) An entity that is capable of reproducing only by infecting a bacterial or eukaryotic cell. Viruses are incapable of autonomous replication and have to use a host cell's translational system. They consist of a nucleic acid molecule and protein coat. The genetic material of a virus may be DNA or RNA. If it is RNA, it will have to be converted to DNA first by the reverse transcriptase enzyme encoded by the viral nucleic acid. These viruses are called retrovirus.

V-J joining. The joining of a variable (V) gene segment and a joining (J) gene segment in the first step of the formation of a functioning immunoglobulin gene.

W

W, Z chromosomes. Sex chromosomes in species (like snakes, birds, moths) where the female is the **heterogametic** sex (WZ).

Wahlund effect. (1) A subdivided population contains fewer heterozygotes than predicted despite the fact that all subdivisions are in Hardy-Weinberg equilibrium. (2) The finding of excess homozygosity in a large sample of population consisted of several subpopulations. It is due to differences in gene frequencies in the subpopulations and purely a mathematical complication.

western blot. A technique for analyzing mixtures of proteins to show the presence, size and abundance of one particular type of protein. Similar to Southern or Northern blotting (see "Blotting"), except that (1) a protein mixture is electrophoresed in an acrylamide gel, and (2) the "probe" is an antibody which recognizes the protein of interest, followed by a radioactive secondary probe (such as 125I-protein A).

western blotting. A technique for identifying a particular protein using antibodies after electrophoretic separation in a gel and transfer to a membrane. See Southern blotting.

wild type. (1) The customary phenotype or standard for comparison. Deviants from this type are said to be

X-inactivation centre (XIC). Locus on the X chromosome in mammals at which inactivation is initiated.

X-ray crystallography. A technique, using X rays, to determine the atomic structure of molecules that have been crystallized. A technique for deducing molecular structure by aiming a beam of X rays at a crystal of the test compound and measuring the scatter of rays.

Y

Y linkage. The inheritance pattern of genes found on the Y chromosome but not on the X (rare).

Y linked. Inheritance pattern of loci located on the Y chromosome only. Also refers to the loci themselves.

YAC. Yeast artificial chromosome. This is a method for cloning very large fragments of DNA. Genomic DNA in fragments of 200-500 kb are linked to sequences which allow them to propagate in yeast as a mini-chromosome (including telomeres, a centromere and an ARS - an autonomous replication sequence). This technique is used to clone large genes and intergenic regions, and for chromosome walking.

yeast artificial chromosome (YAC). (1) Originating from a bacterial plasmid; a YAC contains additionally a yeast centromeric region (CEN); a yeast origin of DNA replication (ARS); and two telomere regions (TEL). YACs are capable of cloning very large pieces of DNA. (2) An artificial chromosome created from DNA, centromere and telomere of yeast chromosomes. Heavily used in cloning of very large genomic fragments.

yeast. The genus Saccharomycetes of the unicellular fungi.

y-junction. The point of active DNA replication where the double helix opens up so that each strand can serve as a template.

Z

Z DNA. A left-handed form of DNA found under physiological conditions in short GC segments that are methylated. It may be important in regulating gene expression in eukaryotes.

zebrafish. A model organism (Danio rerio) to study vertebrate biology, physiology and human disease. Its high fecundity and short generation time make it useful for genetic studies as well. Another useful feature is that their fry are transparent. Hundreds of mutants resembling human diseases have been identified.

ZFY gene. Originally believed to be the human male sex switch gene located on the short arm of the Y chromosome. ZFY stands for zinc finger protein on the Y chromosome.

zinc finger protein. A DNA-binding domain of a protein that has a characteristic pattern of cysteine and histidine residues that complex with zinc ions. This motif occurs in several types of eukaryotic transcription factors.

zinc finger. (1) A protein structural motif common in DNA binding proteins. Four Cys residues are found for each "finger" and one finger can bind a molecule of zinc. A typical configuration is: CysXxxXxxCys—(intervening 12 or so aa's)—CysXxxXxxCys. (2) Configuration of a

DNA-binding protein that resembles a finger with a base, usually cysteines and histidines, binding a zinc ion. Discovered in a transcription factor in Xenopus but present in a large number of different proteins.

zygonema (zygotene stage). The stage of prophase I of meiosis in which synapsis occurs.

zygote. (1) A cell formed by the fusion of sperm and egg. It develops to become an embryo. (2) The unique diploid cell formed by the fusion of two haploid cells (often an egg and a sperm) that will divide mitotically to create a differentiated diploid organism.

zygotic induction. When a prophage is passed into an F-cell during conjugation it may begin vegetative growth. The sudden release of a lysogenic phage from an Hfr chromosome when the prophage enters the F-(minus) cell, and the subsequent lysis of the recipient cell.

zygotic selection. The forces acting to cause differential mortality of an organism at any stage (other than gametes) in its life cycle.